VOLUME ONE HUNDRED AND THREE

INTERNATIONAL REVIEW OF NEUROBIOLOGY

Bioinformatics of Behavior: Part 1

INTERNATIONAL REVIEW OF NEUROBIOLOGY

VOLUME 103

VOLUME ONE HUNDRED AND THREE

INTERNATIONAL REVIEW OF NEUROBIOLOGY

Bioinformatics of Behavior: Part 1

Edited by

ELISSA J. CHESLER
The Jackson Laboratory,
Bar Harbor, Maine, USA

MELISSA A. HAENDEL
Oregon Health & Science University,
Portland, Oregon, USA

ELSEVIER

AMSTERDAM • BOSTON • HEIDELBERG • LONDON
NEW YORK • OXFORD • PARIS • SAN DIEGO
SAN FRANCISCO • SINGAPORE • SYDNEY • TOKYO
Academic Press is an imprint of Elsevier

Academic Press is an imprint of Elsevier
32 Jamestown Road, London NW1 7BY, UK
Radarweg 29, PO Box 211, 1000 AE Amsterdam, The Netherlands
The Boulevard, Langford Lane, Kidlington, Oxford, OX51GB, UK
225 Wyman Street, Waltham, MA 02451, USA
525 B Street, Suite 1900, San Diego, CA 92101-4495, USA

First edition 2012

ISBN: 978-0-12-388408-4
ISSN: 0074-7742

For information on all Academic Press publications
visit our website at store.elsevier.com

Printed and bound in USA
12 13 14 15 11 10 9 8 7 6 5 4 3 2 1

CONTENTS

CONTRIBUTORS

Kyle H. Ambert
Department of Medical Informatics & Clinical Epidemiology, Oregon Health & Science University, Portland, OR, USA

Vadim Astakhov
Department of Neurosciences and Center for Research in Biological Systems, University of California, San Diego, California, USA

Erich J. Baker
Department of Computer Science, Baylor University, Waco, Texas, USA

Anita Bandrowski
Department of Neurosciences and Center for Research in Biological Systems, University of California, San Diego, California, USA

Jonathan Cachat
Department of Neurosciences and Center for Research in Biological Systems, University of California, San Diego, California, USA

Elissa J. Chesler
The Jackson Laboratory, Bar Harbor, Maine, USA

Aaron M. Cohen
Department of Medical Informatics & Clinical Epidemiology, Oregon Health & Science University, Portland, OR, USA

Georgios V. Gkoutos
Department of Genetics, University of Cambridge, Cambridge, UK, and Department of Computer Science, University of Aberystwyth, Old College, Aberystwyth, UK

Jeffery S. Grethe
Department of Neurosciences and Center for Research in Biological Systems, University of California, San Diego, California, USA

Amarnath Gupta
Department of Neurosciences and Center for Research in Biological Systems, University of California, San Diego, California, USA

Melissa A. Haendel
Oregon Health & Science University, Portland, Oregon, USA

Janna Hastings
Swiss Center for Affective Sciences, University of Geneva, Geneva, Switzerland, and Cheminformatics and Metabolism, European Bioinformatics Institute, Cambridge, UK

Robert Hoehndorf
Department of Genetics, University of Cambridge, Cambridge, UK

Fahim Imam
Department of Neurosciences and Center for Research in Biological Systems, University of California, San Diego, California, USA

Stephen D. Larson
Department of Neurosciences and Center for Research in Biological Systems, University of California, San Diego, California, USA

Maryann E. Martone
Department of Neurosciences and Center for Research in Biological Systems, University of California, San Diego, California, USA

Scott F. Saccone
Department of Psychiatry, Washington University, Saint Louis, Missouri, USA

Paul N. Schofield
Department of Physiology, Development and Neuroscience, Downing Street, Cambridge CB2 3EG, UK

Stefan Schulz
Institute for Medical Informatics, Statistics and Documentation, Medical University of Graz, Graz, Austria

PREFACE

The field of bioinformatics has rapidly evolved and has changed the practice of biology in innumerable ways. The impact of modern practices in data management, high-throughput quantitation, semantic data integration, image analysis, text processing, and genomics has changed the scale and breadth of investigation in all areas of biology. These volumes focus on the unique challenges and opportunities of bioinformatics strategies in behavioral science. The first focuses primarily on biological databases and data integration. The second focuses primarily on functional genomics and model organism studies of behavior. Both contain a mixture of theoretical and applied aspects of bioinformatics.

In the development of this work, we faced two major challenges—the tremendous breadth and interdisciplinary nature of bioinformatics, and the highly dynamic nature of the resources developed by bioinformaticians as they leverage new technologies and new points of articulation of neurobehavioral data. We therefore understood that neither could this collection be sufficiently comprehensive nor would the details of various system operations remain static. We chose representative topics and concepts that highlight the issues faced by data analysts, systems designers, and researchers in the behavioral sciences. While the precise resources and applications may change rapidly, we hope that readers gain insight into the strategies, concepts, and considerations in the design, development, and use of these systems in behavioral neurobiology.

For informaticist working with behavioral scientists, we hope our collection highlights the complexities of behavioral data and the unique issues that one may face in trying to define and characterize behavior, an act that may at first appear akin to nailing pudding to a wall. For the behavioral scientist, we hope that we have provided a description of the tools and approaches of the informaticist, whose focus on constrained relations, definitions, and data structures may at first seem utterly Kafkaesque. However, a critical synthesis of these sciences may lead to tremendous advances in developing systems tailored to the complexity of behavior, which may in truth be no less complex than any other biological function. We hope that advances in behavioral bioinformatics and the content herein will engage a new cohort of behavioral deconstructionists, leading us to a new understanding of the biological basis of behavior.

Elissa J. Chesler
Melissa A. Haendel

CHAPTER ONE

Lost and Found in Behavioral Informatics

Melissa A. Haendel[*,1], **Elissa J. Chesler**[†]
[*]Oregon Health & Science University, Portland, Oregon, USA
[†]The Jackson Laboratory, Bar Harbor, Maine, USA
[1]Corresponding author: e-mail address: haendel@ohsu.edu

Contents

Abstract

From early anatomical lesion studies to the molecular and cellular methods of today, a wealth of technologies have provided increasingly sophisticated strategies for identifying and characterizing the biological basis of behaviors. Bioinformatics is a growing discipline that has emerged from the practical needs of modern biology, and the history of systematics and ontology in data integration and scientific knowledge construction. This revolution in biology has resulted in a capability to couple the rich molecular, anatomical, and psychological assays with advances in data dissemination and integration. However, behavioral science poses unique challenges for biology and medicine, and many unique resources have been developed to take advantage of the strategies and technologies of an informatics approach. The collective developments of this diverse and interdisciplinary field span the fundamentals of database development and data integration, ontology development, text mining, genetics, genomics, high-throughput analytics, image analysis and archiving, and numerous others. For the behavioral sciences, this provides a fundamental shift in our ability to associate and dissociate behavioral processes and relate biological and behavioral entities, thereby pinpointing the biological basis of behavior.

International Review of Neurobiology, Volume 103
ISSN 0074-7742
http://dx.doi.org/10.1016/B978-0-12-388408-4.00001-0

1. INTRODUCTION

Genetics and genomics may have given rise to the earliest efforts in what most people think of when they hear "bioinformatics." Bioinformatics is a rapidly evolving interdisciplinary field at the intersection of computer science, database design, molecular science, and functional biology. Though initially focused on storage and analysis of an ever-expanding wealth of DNA sequence data, modern approaches are increasingly focused on relating such molecular entities to organismal function. The application of high-throughput assessment of the role of biological molecules in behavioral processes has given rise to a wealth of data. In human genetics, the major challenge is to find the actual genetic variants responsible for behavioral disorders. Today, bioinformatics provides a diverse array of innovative tools and applications that can be harnessed to further our understanding of the biological underpinnings of human disease.

Behavioral neuroscience provides particular opportunities and challenges for bioinformatics. Behavioral neuroscience has always been a unique discipline—extending and applying advanced methods in many aspects of biology to deciphering abstract behavioral processes. A major challenge has been to describe, define, and discriminate among these abstract behavioral processes, in large part by distinguishing among the biological mechanisms of unique but not entirely discrete, entities of behavior. It is quite apparent that understanding the complexity of neurobiology and behavior requires integration of data across diverse biological systems, types of data, and levels of scale. Bioinformatics is an interdisciplinary field, comprised of people who often have knowledge of computer science and biology, as well as information science and knowledge engineering. Here, we describe how these disciplines can be brought to bear to understand the biological basis of organismal behavior.

2. MAJOR THEMES IN THE BIOINFORMATICS OF BEHAVIOR

2.1. Standardizing data

One of the issues that science faces today is that while we have a wealth of literature from which to draw our conclusions and develop new hypotheses, we do not uniquely identify enough aspects of the research to enable adequate research reproducibility. In other words, one's ability to reproduce the findings described in the literature is hampered by a lack of specificity when referencing the organisms, genes, phenotypes, etc. This problem of experimental reproducibility was the focus of a recent paper by researchers at Amgen in the journal *Nature*, who found that only 11% of the academic research in the literature was reproducible by their groups (Begley & Ellis, 2012). Of course, experimental design, experimental bias, and statistical power may also impact the reproducibility of science—these are features of the scientific method itself that all scientists strive to improve upon. Given that biotech companies have a financial responsibility to select, reproduce, and further develop research around public findings, it is particularly worrisome to think about how the public domain is performing in these respects. Private companies *must* keep explicit track of every aspect of their research, for financial and legal reasons, but also for scientific ones for maximum potential gain. Unfortunately, this philosophy is still young in the public sector, and no informatics volume should be without a short lecture on the unique identification of research entities. The reality is you cannot compute on things for which you have insufficient information on which to perform those computations. Further, education in scientific design that in particular has a focus on informatics methods during the course of research can help and is one of the goals of this volume.

In addition to the lack of unique reference to the processes and entities of research, there are also numerous examples where scientific data or claims are later found to be erroneous or inconsistent. Figure 1.1 shows the results of a search in PubMed for "erratum," which retrieves 4803 results. On the right upper corner, you can see the results by year—there appears to be a trend toward publishing erratum from the mid 1980s to the mid 1990s (one might hypothesize bad music in the labs). One example is a recent analysis of the literature with respect to identification of brain volume differences in various mental disorders (Ioannidis, 2011), which found a statistically significant implausible literature bias toward increased brain volumes. Another example relating to the insufficient reference to animal model experiments comes from statistical analysis of over 50 ALS model SOD1^{G93A} mouse studies. This analysis examined studies on the effects of various drugs and showed how specific biological variables should be controlled for when designing and interpreting efficacy studies, as most drug efficacy conclusions were not reproducible (Scott et al., 2008). There are multiple reasons for the deficiencies pointed out in these meta-analyses, but it is clear that a unique indication of specific organisms, assays, brain regions, behavior being assayed, specific clinical instrument, or other clinical criteria being used

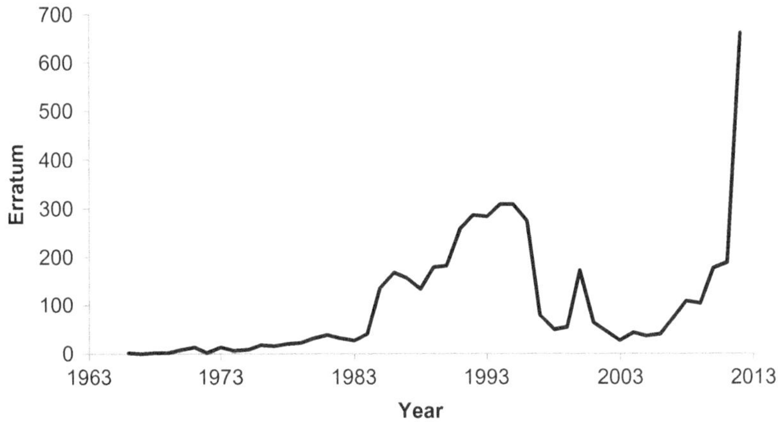

Figure 1.1 A PubMed query for "erratum" produces 4800 results, with the highest rates between 1985 and 1996, with a spike in 2012.

for assessment, etc., could greatly facilitate data aggregation and resolve conflicting claims in the literature, highlighted in this recent editorial (The '3Is' of animal experimentation, 2012).

In particular, and most easily corrected, improper or missing reference to research resources such as antibodies and model organisms, makes it difficult to reproduce scientific evidence or resolve conflicting data. This is a very significant issue in science today, and numerous initiatives, projects, and working groups have been working to address various aspects of the problem (e.g., http://biosharing.org/, http://scientificdatasharing.com/, http://www.data.gov/, and http://datadryad.org/) including recent Requests for Information from the US Office of Science and Technology Policy and the National Institutes of Health (NIH). Potentially even more informative, are recent innovative efforts to analyze the propagation and evolution of assertions in the literature (see Greenberg, 2009), and recent review (Evans & Rzhetsky, 2011), which in the end will rely on the specific reference to research entities to clarify and elucidate scientific facts from fiction. Because such issues have recently come into the limelight, institutional libraries are now performing landscape analyses regarding data management needs (see the Research Data Stewardship at UNC report, 2012) and hiring in-house data management specialists to help support their local research communities. There is a clear need for every scientist to understand how to manage, navigate, and curate their own data (Haendel, Vasilevsky, & Wirz, 2012). The first step toward doing so is to uniquely identify aspects of the scientific process for which standards exist and to document and contribute to those for which no standards exist. Behavioral neuroscience is no exception.

Information science is a field that address the organization, storage, naming, classification, and reasoning over various pieces of information. From computer science, we have the data structures, hardware systems, visualization technologies, all of which are at times the enabling technologies, but at times constraining systems in the execution of modern biology. Understanding how these systems function and what their requirements are is critical to understanding the strategies and approaches used in bioinformatics of behavior. Biological databases for behavioral neurobiology are described in Chapter 2. The chapter by Cachat et al. highlights efforts to integrate data in the neurosciences in the Neuroscience information framework (NIF; Chapter 3). Jay in Volume 104 (Chapter 1) further illustrates what can be done across systems when adequate identifying information is provided from disparate data sources, as exemplified by the GeneWeaver system. Saccone's description of *in silico* integrative genetics—integrating genomic data with genetic studies of human disease in Chapter 7 is another example of the utility of making such data available and uniquely identified. While these chapters specifically focus on data integration, unique identification of data, and making data publicly accessible, there is not a single chapter in these two volumes that do not rely upon unique reference to some biological entity in order to facilitate data capture or analysis in behavioral neuroscience.

2.2. Use of model and not-so-model organisms in the study of behavior

The scientific community has invested heavily in the development of model organisms that potentially recapitulate various aspects of disease. Large genetic screens are performed in organisms such as mice, rats, zebrafish, and *drosophila* to identify new model organisms suitable for the study of various disease attributes. Such model organisms have greatly informed our understanding of human disease and are an essential element in the process of drug development. However, it remains difficult to identify organisms suitable for one's research or assay because information about them is often not readily accessible as per the discussion above. Further, criticisms regarding model organisms abound, because model systems typically do not replicate all aspects of a human disease or disorder. The first problem is simply unique reference to such organisms. Each model organism may represent a genotype, strain, wild-type, background, etc., with one or more sets of identifiers representing that particular organism, or worse, free text labels with no identifiers. It is therefore difficult to identify such resources if they are not consistently referred to in the literature or data sets, in such a way so as to be interoperable with standardized gene representations or public repositories of organism information.

To reduce the variability in the way in which we reference and capture information about model organisms and to collate and make such information publicly available, NIH funds research in 13 "official" model organisms (see http://www.nih.gov/science/models/). Information about these animal models is captured in Model Organism Databases (MODs), described in an overview by Shimamoya (Volume 104, Chapter 2). Extensive curation of the literature and standardization of gene and strain nomenclature for each model organism species are a focus of the MODs. The chapter on informatics resources for mouse (Bult, Volume 104, Chapter 4) focuses specifically on the representation and capture of mouse phenotypes and how they relate to specific assays and genetics in the MOD "Mouse Genome Informatics" or MGI, and in the associated "Mouse Phenome Database" or MPD. New model organism communities wishing to begin similar efforts to those described above may take advantage of the Generic Model Organism Database (GMOD; http://gmod.org/wiki/Main_Page; Mungall & Emmert, 2007) infrastructure to record new model organism data so as to be consistent with other existing MODs. A large, consortial effort to begin to record and study a range of phenotypes, including behavioral assays, across mice mutants for every gene in the genome are described by Morgan et al. in Chapter 3 (Volume 104). Such high-throughput screening techniques will undoubtedly uncover exciting relationships between genetics, behavioral outcomes, and undiscovered phenotypic correlations. As we seek to integrate a multiplicity of data about behavior and how it relates to genetics, genomics, environment, disease and disorders, it will become critical to a developing bioinformatician in behavioral neuroscience to understand and be able to navigate the content of these MOD and MOD-related resources.

Behavioral assays, such as the well-known radial arm maze used in rodents, to memory tasks and motor activity assays in fish, to addictive behavior analyses in fruit flies, have been developed to investigate nervous system function and behavioral development in organisms as diverse as *drosophila* (van Swinderen & Brembs, 2010), *xenopus* (Blackiston & Levin, 2012), *Caenorhabditis elegans* (Kaplan et al., 2012), and zebrafish (Colwill & Creton, 2011), in addition to the more commonly leveraged mammals (discussed in detail in Volume 104, Chapters 2–4), and non-model organisms such as crustaceans (Fernandez De Miguel, Cohen, Zamora, & Arechiga, 1989), planaria (Humphries, 1961; Lee, 1963), and amphibians. For instance, Mathis, Ferrari, Windel, Messier, and Chivers (2008) showed how embryonic exposure to predators in different amphibians alters post-hatching behavior and habitat selection. Assays such as these highlight how behavior is itself a developmental process that happens concurrently with nervous system development and can be used to investigate changes in gene expression as it relates to learning, memory, and behavior, as well as epigenetic factors. For example, alcohol-treated zebrafish have been used as models of fetal alcohol syndrome and show deficiencies in feeding site memory tasks following ethanol exposure earlier in life (Carvan, Loucks, Weber, & Williams, 2004). Deficiencies in swimming activity persist in juveniles that are developmentally exposed to ethanol, an effect mediated in part by miRNAs identified in gene expression profiling studies that also influence brain morphogenesis when knocked down (Tal et al., 2012). Fruit flies have been shown to have an increased preference for ethanol following sexual deprivation, an behavior that appears to be mediated by neuropeptide F (NPF; the mammalian homolog of neuropeptide Y) linking social experience, NPF, and ethanol-related behaviors (Shohat-Ophir, Kaun, Azanchi, & Heberlein, 2012). The development and use of high-throughput systems for a diversity of organisms and behavioral assays have recently been reviewed in Blackiston, Shomrat, Nicolas, Granata, and Levin (2010). High-throughput behavioral analysis of mutant or drug screens in is routinely performed in a variety of organisms (Chan, Inan, Bhattacharya, & Marcu, 2012; Chronis, Zimmer, & Bargmann, 2007; Creton, 2009; Cronin et al., 2005; Kokel et al., 2010). Standardized representation of such behavioral assays, similar to other types of biological assays (see Brinkman et al., 2010; Shimoyama et al., 2012), can enable better query for behavioral phenotypes across data sets.

Increasingly MODs make use of tools that incorporate mapping to other species, and many tools and approaches have been developed to perform global analysis of the data they contain (see Volume 104, Chapters 2–4). Model organisms are a powerful resource for the discovery of genes, networks, and pathways underlying behavioral variation, but leave behavioral scientists, particularly those hoping to address human conditions, with a fundamental challenge of extrapolation. A major impediment in bioinformatics is to compare biological substrates across species. This can be done at several levels, the most basic being through homology of genes and gene products.

Compelling success stories have revealed the shared role of homologous genes across species for numerous behaviors. However, in many cases, the precise molecular players may differ across species, though the net result may be conserved. Strategies that attempt to match convergent pathway utilization (rather than specific genes within the pathways) across species may therefore be a more effective solution to comparative functional genomics. The tremendous diversity of data, experiment types, and species applied to behavioral science create numerous challenges for those wishing to employ these types of techniques in their own labs (Volume 104, Chapter 1), and a variety of new software has been developed to address these issues (see the list of links at the end of Volume 104).

One common criticism of using animal models for the study of disease is that the models typically only recapitulate some portion of the disease phenotypes, the observable outcomes of the synergy between gene expression and environmental factors over time. Classification of animal models of disease based on assertions that a given organism is a "model of disease X" does not solve the issue because in addition to the model not recapitulating all aspects of the disease, the specific aspects that relate the model to the disease are not usually indicated. These assertions between a model organism and a disease also specifically give a misleading impression for behavioral disorders, because such disorders largely encompass a heterogeneous group of endophenotypes, or atomic cognitive, electrophysiological, or neuroimaging measures (originally coined by Gottesman and Shields (1973), also see Volume 104, Chapter 8). Conversely, analysis of model systems tends to focus on structures or outcomes that were identified as being important in a disease rather than providing a global characterization, limiting their utility in particular for behavioral neuroscience. As a result, many potential models for behavioral disease or disorder have likely gone unidentified. New screens, such as that described in mouse (see Volume 104, Chapter 4) attempt to provide a more global overview of mutant phenotyping that includes behavioral assessment. While such advancements and considerations will help increase our knowledge of behavioral phenotypes, it is clear that a more powerful and granular system is needed to describe and query models of behavioral dysfunction.

Of course, given the decreasing cost of genome sequencing, it now makes sense to consider organisms other than those traditionally thought of as model organisms when attempting to link genomics to behavioral outcomes. Such systems can in fact inform our understanding of behavior and behavioral dysfunction. For instance, Dr. Smith has been working on the genetic basis of Williams–Beuren Syndrome, a disorder that presents with over-social or gregarious behavior – in ants. People who have the disorder have been shown to be missing a chromosomal region containing 26 genes on Chromosome 7, more than half of these genes have orthologs in ants (Gadau et al., 2012). Since the contribution of these genes to behavior is unknown, Smith and colleagues have altered the expression of these genes to determine if such changes alter the ants' social behavior (personal communication, and also see San Francisco State University Newsletter, 2012). Why ants? Ants exhibit complex social behavior in ways that are surprisingly similar to humans. Such genomic comparisons as described by Gadau will enable an analysis of different species of ants to determine if they have evolved different "sociogenomes."

Ants are not the only non-model organisms that exhibit complex behaviors. The sophisticated courtship and other behaviors of spiders have been studied and recorded using ontologies (Arachnolignua oral presentation at iEvoBio, 2012), which are semantic representations that enable inference based on the logical definitions of the vocabulary classes and properties between classes. An ontology, used in combination with logical inference tools called reasoners (e.g., see Kazakov & Krötzsch, 2012; Sirin, Parsia, Grau, Kalyanpur, & Katz, 2007), can assist in answering queries and grouping and comparing data by leveraging the logical relationships between the concepts that comprise the ontology. "Arachnolingua" is a knowledgebase of published descriptions of behaviors performed by spider species and is a resource of non-model organism behavior used in part to test and extend the NeuroBehavior Ontology (see Chapter 4). In this way, even without genomics we can learn to better classify the atomic phenotypes, or endophenotypes of behavior, and thereby apply such informatics to the representation of human behavior and relate it to the wealth of omics data (as per Saccone, Chapter 7 and Jay, Volume 104, Chapter 1). Further, behavioral phenotypes inhere in populations as well as individuals, such as bird nest-building behaviors, and in response to predation or environmental factors; such suites of population-level behaviors are referred to as behavioral syndromes (Sih, Bell, & Johnson, 2004). It will be interesting to see how efforts such as those described in the chapter on representation of clinical behavioral data, such as cognition, perception, and emotion, by Hastings and Schulz (Chapter 5) will enable comparison and integration of behavioral phenotypes in other organisms. A recent consideration of executive function as it relates to brain cytoarchitectural evolution highlights the utility of structure–function correlation (Bilder, 2012). Perhaps the day is not far away when the way in which we classify a spider behavior may inform an understanding of behavioral phenotypic profiles in humans and subsequent clinical decision-making processes.

2.3. Speaking the same behavioral language

Mapping the activity of biological entities onto behavior requires just that—mapping. The heroic Allan Brain Atlas project aims to physically localize gene expression in the developing and adult mouse brain (described in Volume 104, Chapter 7), and thereby relate structure to function. High-throughput genomic studies summarized in Volume 104, Chapter 5 provide whole genome quantitation of the role of gene products in behavioral processes and have advanced from simple studies of differential expression to complex network analyses aimed at reconstructing the effector pathways of behavioral change. A systems biological approach to mouse genetics is used in the GeneNetwork system, described by Williams and Mulligan in Chapter 6, Volume 104, which enables the discovery and testing for associations between differences in DNA sequence and behavioral variation. Functional electromagnetics is a strategy described by Frishkoff et al. (2007, 2011) wherein by classifying brain activity patterns as they relate to behavioral tasks across studies, one can start to conceive of how such activity patterns might be the missing link between behavior and alterations in gene expression.

A significant barrier to query between human and model systems is due to the difference in terminology used to describe them. Each organism has its own vocabulary for describing the phenotypic consequences of mutation, which is particularly evident when trying to compare clinical and research data about organisms. Even researchers who study the same organism may have significant communication difficulties. Neurophysiologists, for example, usually describe their data relative to a functionally specified brain region such as the "primary auditory cortex"; neuroanatomists describe the same area as a Brodmann area 41 or 42, and the latter does not necessarily spatially or conceptually overlap completely with the former. Such differences in terminology make it difficult for automated agents like text-mining tools to draw comparisons among human disorders and relevant animal models. Disease and phenotype descriptions are often recorded as free text, and although descriptive, natural language remains difficult to computationally compare. However, if an organism's behavioral phenotypes were semantically linked to diseases, genes, phenotypes, expression profiles, etc., their relevance to a particular area of research would potentially be revealed. Despite this well-recognized problem, the bioinformatics tools required to facilitate identification of models of disease have been lacking because the relationship between gene and disease (Strohman, 2002) and between model system and disease phenotypes (Houle, Govindaraju, & Omholt, 2010) is not straightforward. This is especially a problem for behavioral phenotypes, which are often not well described.

For behavioral disorders in which the genetic basis is unknown or there is no genetic basis, the identification of sequence orthologs does not help identify models of disease. What is needed is a computational approach to determine similarity between phenotypes to identify candidate models. When phenotype descriptions are captured using an ontology, algorithms can be written to compare phenotypes computationally. Ontologies and data standards have been used to meaningfully relate gene function, expression, proteins, and more (examples in Andronis, Sharma, Virvilis, Deftereos, & Persidis, 2011; Brochhausen et al., 2011; Consortium, 2009; Field et al., 2009) and a number of relevant efforts have utilized ontologies for mining phenotypes. Schlicker et al. (Schlicker & Albrecht, 2008; Schlicker, Lengauer, & Albrecht, 2010) have analyzed phenotypic profiles using the species-neutral Gene Ontology and a specific list of proteins and gene–disease associations from the Online Mendelian Inheritance in Man. PhenomicDB (Groth et al., 2007, 2010) is a cross-species resource that aggregates ontology annotations from diverse resources and mines free-text phenotypes to provide "phenoclusters" of phenotype-related genes. While these methods are very useful for comparing phenotypes based on gene orthology or gene annotations, they do not enable discovery of similar phenotypes based solely on phenotype descriptions. The emphasis has on identification of "responsible" genes, rather than a focus on the phenotype description itself makes such approaches more limited for the analysis of behavioral phenotypes. In a recent issue of the journal *Human Mutation* specifically on phenotype analysis, Dr. Robinson defines deep phenotyping as "as the precise and comprehensive analysis of phenotypic abnormalities in which the individual components of the phenotype are observed and described" (2012). Ontological annotation of behavioral diseases and phenotypes can provide this "deep phenotyping" and thereby enable computational comparison of phenotypes across species in the absence of genetic information.

One of the challenges in comparing phenotypes or gene expression across species is the lack of a mechanism to traverse anatomical structures.

Computers are not aware that the human auditory cortex may be related in some fashion to the zebrafish pallial amygdala (Mueller, 2012) because they do not know that the two structures are both part of the brain in those species, nor even that zebrafish brain is related to the human brain. A new ontology has been created that attempts to address this issue, Uberon, which classifies anatomical structures via a variety of axes such as structure, function, and development, and relates them back to the species-specific anatomies for cross-species inference (Mungall, Torniai, Gkoutos, Lewis, & Haendel, 2012). Specifically, Uberon is being used to enhance interoperability with ontologies such as the Mammalian Phenotype Ontology (Smith & Eppig, 2009; Smith, Goldsmith, & Eppig, 2005; see also Volume 104, Chapters 2–4) and the Human Phenotype Ontology (Robinson & Mundlos, 2010; Robinson et al., 2008), allowing them to be integrated with other phenotype data (Gkoutos et al., 2009; Hancock et al., 2009; Hoehndorf, Schofield, & Gkoutos, 2011; Kohler, Doelken, Rath, Ayme, & Robinson, 2012; Mungall et al., 2010; Washington et al., 2009). Recently, a neurodegenerative disease phenotype knowledgebase called PKB (Maynard, Mungall, Lewis, Imam, & Martone, 2012) has been constructed that utilizes the NIF Standard (Chapter 3) modular collection of ontologies (Bug et al., 2008; Imam et al., 2012) to represent a range of human diseases and animal models spanning multiple anatomical scales, from the molecular and subcellular up to the organismal. This illustrates significant progress toward computability of phenotypes at different levels of anatomical granularity and use of many different vocabularies to express the phenotypes, which will be critical for the investigation of behavior.

Another approach to querying for similar phenotypes combines orthology and gene–phenotype ontology associations was used to generate "phenolog" hypotheses, non-obvious linkages between human diseases and asserted phenotypes from MODs such as mouse, worm, yeast, and plant (McGary et al., 2010). This approach can be extended to suggest new models, based on the presence of orthologous genes inside a phenolog cluster. Related approaches make further use of the semantic relations in the data, such as in MouseFinder (Chen et al., 2012). With respect to cognitive phenotypes, some have posed that use of endophenotypes does not improve understanding of the genetic basis of behavioral disorders over syndrome-based associations in GWAS studies (Flint & Munafo, 2007). However, it is clear that representation of such atomic phenotypes furthers our understanding of such disorders and fosters communication and integration of data about them. New studies are emerging that are beginning to realize such efforts to "atomize" the phenotypes, represent them using ontologies, and identify new gene candidates based on atomic phenotypes. Meehan et al. (2011) identified candidate genes based on analysis of the intersection of rare CNVs implicated in autism and mammalian phenotype ontology annotations to identify mouse models of autism based on human phenotypes. In this way, one can leverage ontologies and in particular endophenotypes or behavioral traits, to enable better use of model organisms in the identification and development diagnostic and therapeutic targets. Similarly, endophenotypes are being leveraged in the GeneNetwork analysis of mouse behavior to identify mouse models of behavioral disorders (see Volume 104, Chapter 6). Efforts such as these will identify those model organism characteristics that share common substrates with psychiatric conditions in people. With the Personal Genome Project (http://www.personalgenomes.org/) aiming to enroll 100,000 informed participants who are willing to share their genome, it may be possible to begin to leverage human behavioral data in phenotype similarity analyses.

The unique challenges in the naming and identification of behaviors have been in part addressed through efforts at developing ontologies and a number of projects aim to develop cognitive ontologies. One such collection of ontologies is being developed collaboratively at the Consortium for Neuropsychiatric Phenomics (www.phenomics.ucla.edu), to enable linking of information about cognitive phenotypes to other biological knowledge (Bilder et al., 2009). Bilder suggests that, for example, "perhaps a stronger genetic association might be found for individuals with poor premorbid social function, gray matter volume reduction, poor working memory, and negative symptoms, than could be found for any one of these alone." To paraphrase Bilder, the suggestion is that if one more adequately defines phenotypes, then one may leverage the increased numbers of paths that relate genotype to phenotype. Several chapters in this volume discuss the development of ontologies for the classification of behavioral traits, which can be leveraged to relate behavior to numerous other data facets. Gkoutos describes the Neuro-Behavior ontology, which aims to standardize representation of behavior across species including human disorders (Chapter 4). Hastings and Schulz describe vocabularies used for clinical classification of behavioral dysfunction, such as SNOMED and DSM-IV, and how they relate to more formal ontology efforts to represent behavior (Chapter 5). These efforts have the end-goal to anchor measurements to a classification of the kinds of cognitive entities that exist, such as "short-term memory" or "sadness." Such cognitive concepts are of obvious difficulty to define, and attempts to reconcile community differences in defining such a classification are being addressed as part of the Cognitive Atlas project (www.cognitiveatlas.org; Poldrack et al., 2011), wherein such conflicts in ontological classification can be resolved with empirical evidence. Despite these hurdles, we can leverage various clinical instruments to aid in the representation and definition of cognitive classification, and thereby gain the power of inference to relate cognition to genetics and brain functioning. As a case in point, Frishkoff et al. (2011) describes the development of a common framework for labeling and classifying neurodynamic patterns in order to compare diverse study contexts and data from different methodologies. The MODs and other related databases all leverage such ontologies (every chapter in this book mentions use of ontologies or data annotated with ontologies in some fashion). However, given that most researchers and clinicians do not walk around with an ontology in their back pocket (ontology-driven tools are unfortunately not yet in common use in the laboratories), text-mining and entity extraction using ontologies can be a good mechanism to extract and relate behavioral data from the literature or other text sources such as electronic health records. Ambert and Cohen describe strategies to extract information from the large volumes of literature produced every year by legions of aspiring young scientists and elder statesmen (Chapter 6). Similarly, the CureHunter system (www.curehunter.com) can interpret the biomedical literature to identify candidate drugs for specific diseases, including a wealth of behaviorally relevant disorders. Can the modern scientist master this literature, or is a computable knowledge framework a critical step beyond the Guttenberg system of knowledge dissemination? Text-mining used in combination with ontologies holds great promise for navigating and inferring new hypotheses from this onslaught of information.

3. FURTHER WORDS

There are numerous methods to analyze behaviorally relevant data, many of which are described herein, and it is the intersection of such methods that we may find to be most fruitful to shed light on the biological basis of behavior. There are potentially innumerable and elusive reasons for this, only some of which are that behavior and assays to measure it are often poorly defined, behavior is the culmination of biological activity at different levels of granularity in time and space, behavior is often affected by numerous genetic and epigenetic mechanisms, and possibly even the fact that humans don't make very good model organisms. How can one overcome such obstacles? Learning to standardize data, adopt nomenclature conventions, and make research database savvy and database enabled is a key to the modern execution of research in biology. It enables a wide audience to operate rapidly on research results, and fosters tacit collaboration. Traversing animal models and integrating data can place individual findings in a better context and provide a global framework for the acquisition and aggregation of knowledge about organismal behavior. Due to the near impossibility of mastering the entire literature in one's field, such indexing is proving critical; though we contend (and reassure the neuroscientist) that this may not yet or ever replace the depth of description and interpretation in the primary literature. Developing an appreciation and familiarity with resources and techniques will enhance even the seemingly least informatics oriented research efforts. We hope this volume provides behavioral neuroscientists an orientation and introduction to some of the critical issues and areas of development in the field.

REFERENCES

Andronis, C., Sharma, A., Virvilis, V., Deftereos, S., & Persidis, A. (2011). Literature mining, ontologies and information visualization for drug repurposing. *Briefings in Bioinformatics*, *12*, 357–368.

Arachnolignua oral presentation at iEvoBio. (2012). http://www.slideshare.net/pmidford/ievobio-2012-lightning-talk-arachnolingua. Accessed 15/08/12.

Begley, C. G., & Ellis, L. M. (2012). Drug development: Raise standards for preclinical cancer research. *Nature*, *483*, 531–533.

Bilder, R. M. (2012). Executive control: Balancing stability and flexibility via the duality of evolutionary neuroanatomical trends. *Dialogues in Clinical Neuroscience*, *14*, 39–47.

Bilder, R. M., Sabb, F. W., Parker, D. S., Kalar, D., Chu, W. W., Fox, J., et al. (2009). Cognitive ontologies for neuropsychiatric phenomics research. *Cognitive Neuropsychiatry*, *14*, 419–450.

Blackiston, D. J., & Levin, M. (2012). Aversive training methods in Xenopus laevis: General principles. *Cold Spring Harbor Protocols*. http://dx.doi.org/10.1101/pdb.top068338.

Blackiston, D., Shomrat, T., Nicolas, C. L., Granata, C., & Levin, M. (2010). A second-generation device for automated training and quantitative behavior analyses of molecularly-tractable model organisms. *PloS One*, *5*, e14370.

Brinkman, R. R., Courtot, M., Derom, D., Fostel, J. M., He, Y., Lord, P., et al. (2010). Modeling biomedical experimental processes with OBI. *Journal of Biomedical Semantics*, *1*(Suppl. 1), S7.

Brochhausen, M., Spear, A. D., Cocos, C., Weiler, G., Martin, L., Anguita, A., et al. (2011). The ACGT Master Ontology and its applications—Towards an ontology-driven cancer research and management system. *Journal of Biomedical Informatics*, *44*, 8–25.

Bug, W. J., Ascoli, G. A., Grethe, J. S., Gupta, A., Fennema-Notestine, C., Laird, A. R., et al. (2008). The NIFSTD and BIRNLex vocabularies: Building comprehensive ontologies for neuroscience. *Neuroinformatics*, *6*, 175–194.

Carvan, M. J., 3rd, Loucks, E., Weber, D. N., & Williams, F. E. (2004). Ethanol effects on the developing zebrafish: Neurobehavior and skeletal morphogenesis. *Neurotoxicology and Teratology*, *26*, 757–768.

Chan, K. L., Inan, O., Bhattacharya, S., & Marcu, O. (2012). Estimating the speed of Drosophila locomotion using an automated behavior detection and analysis system. *Fly*, *6*(3), 205–210. http://dx.doi.org/10.4161/fly.20987.

Chen, C. K., Mungall, C. J., Gkoutos, G. V., Doelken, S. C., Kohler, S., Ruef, B. J., et al. (2012). MouseFinder: Candidate disease genes from mouse phenotype data. *Human Mutation*, *33*, 858–866.

Chronis, N., Zimmer, M., & Bargmann, C. I. (2007). Microfluidics for in vivo imaging of neuronal and behavioral activity in Caenorhabditis elegans. *Nature Methods*, *4*, 727–731.

Colwill, R. M., & Creton, R. (2011). Imaging escape and avoidance behavior in zebrafish larvae. *Reviews in the Neurosciences*, *22*, 63–73.

Consortium, R. G. G. O. T. G. O. (2009). The Gene Ontology's Reference Genome Project: A unified framework for functional annotation across species. *PLoS Computational Biology*, *5*, e1000431.

Creton, R. (2009). Automated analysis of behavior in zebrafish larvae. *Behavioural Brain Research*, *203*, 127–136.

Cronin, C. J., Mendel, J. E., Mukhtar, S., Kim, Y. M., Stirbl, R. C., Bruck, J., et al. (2005). An automated system for measuring parameters of nematode sinusoidal movement. *BMC Genetics*, *6*, 5.

Evans, J. A., & Rzhetsky, A. (2011). Advancing science through mining libraries, ontologies, and communities. *The Journal of Biological Chemistry*, *286*, 23659–23666.

Fernandez De Miguel, F., Cohen, J., Zamora, L., & Arechiga, H. (1989). An automated system for detection and analysis of locomotor behavior in crustaceans. *Boletín de Estudios Médicos y Biológicos*, *37*, 71–76.

Field, D., Sansone, S. A., Collis, A., Booth, T., Dukes, P., Gregurick, S. K., et al. (2009). Megascience. Omics data sharing. *Science*, *326*, 234–236.

Flint, J., & Munafo, M. R. (2007). The endophenotype concept in psychiatric genetics. *Psychological Medicine*, *37*, 163–180.

Frishkoff, G. A., Frank, R. M., Rong, J., Dou, D., Dien, J., & Halderman, L. K. (2007). A framework to support automated classification and labeling of brain electromagnetic patterns. *Computational Intelligence and Neuroscience*, *14567*. http://dx.doi.org/10.1155/2007/14567. PMCID: PMC2246027.

Frishkoff, G., Sydes, J., Mueller, K., Frank, R., Curran, T., Connolly, J., et al. (2011). Minimal Information for Neural Electromagnetic Ontologies (MINEMO): A standards-compliant method for analysis and integration of event-related potentials (ERP) data. *Standards in Genomic Sciences*, *5*(2), 211–223.

Gadau, J., Helmkampf, M., Nygaard, S., Roux, J., Simola, D. F., Smith, C. R., et al. (2012). The genomic impact of 100 million years of social evolution in seven ant species. *Trends in Genetics*, *28*, 14–21.

Gkoutos, G. V., Mungall, C., Dolken, S., Ashburner, M., Lewis, S., Hancock, J., et al. (2009). Entity/quality-based logical definitions for the human skeletal phenome using PATO. *Conference Proceedings: ... Annual International Conference of the IEEE Engineering in Medicine and Biology Society. IEEE Engineering in Medicine and Biology Society. Conference*, *2009*, 7069–7072.

Gottesman, I. I., & Shields, J. (1973). Genetic theorizing and schizophrenia. *The British Journal of Psychiatry*, *122*, 15–30.

Greenberg, S. A. (2009). How citation distortions create unfounded authority: Analysis of a citation network. *British Medical Journal*, *339*, b2680. http://dx.doi.org/10.1136/bmj.b2680.

Groth, P., Kalev, I., Kirov, I., Traikov, B., Leser, U., & Weiss, B. (2010). Phenoclustering: Online mining of cross-species phenotypes. *Bioinformatics*, *26*, 1924–1925.

Groth, P., Pavlova, N., Kalev, I., Tonov, S., Georgiev, G., Pohlenz, H. D., et al. (2007). PhenomicDB: A new cross-species genotype/phenotype resource. *Nucleic Acids Research, 35*, D696–D699.

Haendel, M. A., Vasilevsky, N. A., & Wirz, J. A. (2012). Dealing with data: A case study on information and data management literacy. *PLoS Biology, 10*, e1001339.

Hancock, J. M., Mallon, A. M., Beck, T., Gkoutos, G. V., Mungall, C., & Schofield, P. N. (2009). Mouse, man, and meaning: Bridging the semantics of mouse phenotype and human disease. *Mammalian Genome, 20*, 457–461.

Hoehndorf, R., Schofield, P. N., & Gkoutos, G. V. (2011). PhenomeNET: A whole-phenome approach to disease gene discovery. *Nucleic Acids Research, 39*, e119.

Houle, D., Govindaraju, D. R., & Omholt, S. (2010). Phenomics: The next challenge. *Nature Reviews. Genetics, 11*, 855–866.

Humphries, B. (1961). Maze learning in planaria. *Worm Runner's Digest, 3*, 114–115.

Imam, F. T., Larson, S. D., Bandrowski, A., Grethe, J. S., Gupta, A., & Martone, M. E. (2012). Development and use of ontologies inside the neuroscience information framework: A practical approach. *Frontiers in Genetics, 3*, 111.

Ioannidis, J. P. (2011). Excess significance bias in the literature on brain volume abnormalities. *Archives of General Psychiatry, 68*, 773–780.

Kaplan, F., Alborn, H. T., von Reuss, S. H., Ajredini, R., Ali, J. G., Akyazi, F., et al. (2012). Interspecific nematode signals regulate dispersal behavior. *PloS One, 7*, e38735.

Kazakov, Y., Krötzsch, M., & Simančík, F. Elk Reasoner: Architecture and evaluation. In M. Y. Ian Horrocks, & Ernesto Jimenez-Ruiz (Eds.), *Proceedings of the 1st International Workshop on OWL Reasoner, Evaluation* (*ORE-2012*, P10).

Kohler, S., Doelken, S. C., Rath, A., Ayme, S., & Robinson, P. N. (2012). Ontological phenotype standards for neurogenetics. *Human Mutation, 33*, 1333–1339.

Kokel, D., Bryan, J., Laggner, C., White, R., Cheung, C. Y., Mateus, R., et al. (2010). Rapid behavior-based identification of neuroactive small molecules in the zebrafish. *Nature Chemical Biology, 6*, 231–237.

Lee, R. M. (1963). Conditioning of a free operant response in planaria. *Science, 139*, 1048–1049.

Mathis, A., Ferrari, M. C., Windel, N., Messier, F., & Chivers, D. P. (2008). Learning by embryos and the ghost of predation future. *Proceedings of the Royal Society B, 275*, 2603–2607.

Maynard, S., Mungall, C., Lewis, S., Imam, F., & Martone, M. (2012). A knowledge based approach to matching human neurodegenerative disease and animal models. *BMC Bioinformatics*, (in press).

McGary, K. L., Park, T. J., Woods, J. O., Cha, H. J., Wallingford, J. B., & Marcotte, E. M. (2010). Systematic discovery of nonobvious human disease models through orthologous phenotypes. *Proceedings of the National Academy of Sciences of the United States of America, 107*, 6544–6549.

Meehan, T. F., Carr, C. J., Jay, J. J., Bult, C. J., Chesler, E. J., & Blake, J. A. (2011). Autism candidate genes via mouse phenomics. *Journal of Biomedical Informatics, 44*(Suppl. 1), S5–S11.

Mueller, T. (2012). What is the Thalamus in Zebrafish? *Frontiers in Neuroscience, 6*, 64.

Mungall, C. J., & Emmert, D. B. (2007). A Chado case study: An ontology-based modular schema for representing genome-associated biological information. *Bioinformatics, 23*, i337–i346.

Mungall, C. J., Gkoutos, G. V., Smith, C. L., Haendel, M. A., Lewis, S. E., & Ashburner, M. (2010). Integrating phenotype ontologies across multiple species. *Genome Biology, 11*, R2.

Mungall, C. J., Torniai, C., Gkoutos, G. V., Lewis, S. E., & Haendel, M. A. (2012). Uberon, an integrative multi-species anatomy ontology. *Genome Biology, 13*, R5.

Poldrack, R. A., Kittur, A., Kalar, D., Miller, E., Seppa, C., Gil, Y., et al. (2011). The cognitive atlas: Toward a knowledge foundation for cognitive neuroscience. *Frontiers in Neuroinformatics*, *5*, 17.

Robinson, P. N. (2012). Deep phenotyping for precision medicine. *Human Mutation*, *33*, 777–780.

Robinson, P. N., Kohler, S., Bauer, S., Seelow, D., Horn, D., & Mundlos, S. (2008). The Human Phenotype Ontology: A tool for annotating and analyzing human hereditary disease. *American Journal of Human Genetics*, *83*, 610–615.

Robinson, P. N., & Mundlos, S. (2010). The human phenotype ontology. *Clinical Genetics*, *77*, 525–534.

San Francisco State University Newsletter. (2012). http://news.sfsu.edu/ant-genomes-offer-new-ways-explore-social-behavior. Accessed 15/08/12.

Schlicker, A., & Albrecht, M. (2008). FunSimMat: A comprehensive functional similarity database. *Nucleic Acids Research*, *36*, D434–D439.

Schlicker, A., Lengauer, T., & Albrecht, M. (2010). Improving disease gene prioritization using the semantic similarity of Gene Ontology terms. *Bioinformatics*, *26*, i561–i567.

Scott, S., Kranz, J. E., Cole, J., Lincecum, J. M., Thompson, K., Kelly, N., et al. (2008). Design, power, and interpretation of studies in the standard murine model of ALS. *Amyotrophic Lateral Sclerosis*, *9*, 4–15.

Shimoyama, M., Nigam, R., McIntosh, L. S., Nagarajan, R., Rice, T., Rao, D. C., et al. (2012). Three ontologies to define phenotype measurement data. *Frontiers in Genetics*, *3*, 87.

Shohat-Ophir, G., Kaun, K. R., Azanchi, R., & Heberlein, U. (2012). Sexual deprivation increases ethanol intake in Drosophila. *Science*, *335*, 1351–1355.

Sih, A., Bell, A., & Johnson, J. C. (2004). Behavioral syndromes: An ecological and evolutionary overview. *Trends in Ecology & Evolution*, *19*, 372–378.

Sirin, E., Parsia, B., Grau, B. C., Kalyanpur, A., & Katz, Y. (2007). Pellet: A practical OWL-DL reasoner. *Web Semantics: Science, Services and Agents on the World Wide Web*, *5*, 51–53.

Smith, C. L., & Eppig, J. T. (2009). The mammalian phenotype ontology: Enabling robust annotation and comparative analysis. *Wiley Interdisciplinary Reviews. Systems Biology and Medicine*, *1*, 390–399.

Smith, C. L., Goldsmith, C. W., & Eppig, J. T. (2005). The Mammalian Phenotype Ontology as a tool for annotating, analyzing and comparing phenotypic information. *Genome Biology*, *6*, R7.

Strohman, R. (2002). Maneuvering in the complex path from genotype to phenotype. *Science*, *296*, 701–703.

Tal, T. L., Franzosa, J. A., Tilton, S. C., Philbrick, K. A., Iwaniec, U. T., Turner, R. T., et al. (2012). MicroRNAs control neurobehavioral development and function in zebrafish. *The FASEB Journal*, *26*, 1452–1461.

van Swinderen, B., & Brembs, B. (2010). Attention-like deficit and hyperactivity in a Drosophila memory mutant. *The Journal of Neuroscience*, *30*, 1003–1014.

Washington, N. L., Haendel, M. A., Mungall, C. J., Ashburner, M., Westerfield, M., & Lewis, S. E. (2009). Linking human diseases to animal models using ontology-based phenotype annotation. *PLoS Biology*, 7, e1000247.

The '3Is' of animal experimentation (2012). *Nature Genetics*, *44*, 611.

Research Data Stewardship at UNC: Recommendations for Scholarly Practice and Leadership [Online]. http://sils.unc.edu/sites/default/files/general/research/UNC_Research_Data_Stewardship_Report.pdf. Accessed 08/06/2012.

CHAPTER TWO

Biological Databases for Behavioral Neurobiology

Erich J. Baker[1]
Department of Computer Science, Baylor University, Waco, Texas, USA
[1]Corresponding author: e-mail address: erich_baker@baylor.edu

Contents

Abstract

Databases are, at their core, abstractions of data and their intentionally derived relationships. They serve as a central organizing metaphor and repository, supporting or augmenting nearly all bioinformatics. Behavioral domains provide a unique stage for contemporary databases, as research in this area spans diverse data types, locations, and data relationships. This chapter provides foundational information on the diversity and prevalence of databases, how data structures support the various needs of behavioral neuroscience analysis and interpretation. The focus is on the classes of databases, data curation, and advanced applications in bioinformatics using examples largely drawn from research efforts in behavioral neuroscience.

International Review of Neurobiology, Volume 103
ISSN 0074-7742
http://dx.doi.org/10.1016/B978-0-12-388408-4.00002-2

1. INTRODUCTION

It is difficult to imagine modern neuroscience research without the supporting infrastructure provided by bioinformatics databases. Consistent with the broader view of informatics, a bioinformatics renders a formalized representation of information, placing empirical observations within the context of the larger subdiscipline and augmenting the impact of local observations and experimentation. The ultimate goal is to allow other researchers from a variety of tangential disciplines to share a common lexicon and classification framework to bridge the data-mining gap, automating the process of knowledge discovery. With mature bioinformatics, for example, the broad implications of behavioral neuroscience can be measured against the convergent functional genomics of several model organisms, opening up avenues of validation previously hidden behind isolated or contextually limited data. Additionally, in contrast to reductionists views of physical models, there is no true interpretation of biological data (Birney & Clamp, 2004) and well-conceived database implementations can move semi-quantitative phenotypes or behavioral observations toward a more tightly structured quantitative result without limiting the scope of analysis to domains where the researcher has deep knowledge.

Behavioral neuroscience databases are required to harness the rapid and accelerating volume of new data and to integrate an incredibly diverse set of traditional and high-throughput technologies. The latter use of databases is of particular interest as behavioral neuroscience spans countless experimental designs and geographic locations, but suffers from the universal lack of an organic data format. For example, the Society For Neuroscience has 42,000 members (www.sfn.org), working with a variety of model organisms and focused on an innumerable array of differing physiological depth and developmental timescales. Gaining a mastery of a common literature within this diverse group is daunting, but managing the integration of 42,000 individual lab notebooks in countless formats is not feasible. Without a common data format or meaningful translational key, the intractable density of information within individual data silos can paralyze analytics, causing researchers to shift focus away from the painful difficulty of knowledge discovery within disassociated data and focus on previously explored areas where data types and structures have been well-documented.

Modern open-source database management systems (DBMS) are used by bioinformatics specialists to mediate potential information bottlenecks. Biological databases serve to shift the burden of data management from the researcher onto a generalizable platform, effectively placing information in a layer that performs local information management duties while making itself transparently accessible to analysis tools and other databases (Fig. 2.1). An interesting consequence of database effectiveness and interaction transparency is that researchers have become desensitized to their deep complexities. There is often a failure to recognize the intimate relationship between types of databases, their intended use, and the landscape and provenance of the underlying data. In behavioral neuroscience research, the depth of these relationships is uniquely important because of the underlying breadth of subdomains, the interaction of vastly arrayed qualitative and quantitative data types, and layers of non-overlapping and often ambiguous semantics ranging from molecular to behavioral observations.

This survey of the types and scope of databases useful to behavioral neuroscience illustrates the connections between the varying types of underlying data and the purpose of the database. While there certainly is no singular biological database model that defines the entire granularity implicit within the domain, there does exist an emerging understanding of the opportunities and limitations of neuroscience related biological databases.

2. NEUROSCIENCE DATABASES

Researchers interested in understanding, collating, and analyzing the information of neuroscience have numerous hurdles. From a practical perspective, within the biological database community there is a vacillation between infrastructure building and scholarship, creating competing incentives for finding publishable hypotheses within the tangle of existing databases and the creation of new databases (Altman, 2004). As a result, many life science databases in general and behavioral neuroscience databases in particular have grown out of a single research lab to mediate a particular tactical need. For example, neuroscience databases and data management tools include those seeking to manage transcriptional data (Shepherd et al., 1998), complex images such as fMRI scans (Marcus et al., 2007), laboratory information management systems (LIMS) and data management (Baker, Galloway, Jackson, Schmoyer, & Snoddy, 2004), formal collaborations and federated repositories (Gardner et al., 2008), publication data (Ruttenberg, Rees, Samwald, & Marshall, 2009), protein interaction (Colland et al., 2004; Shoemaker et al., 2012)

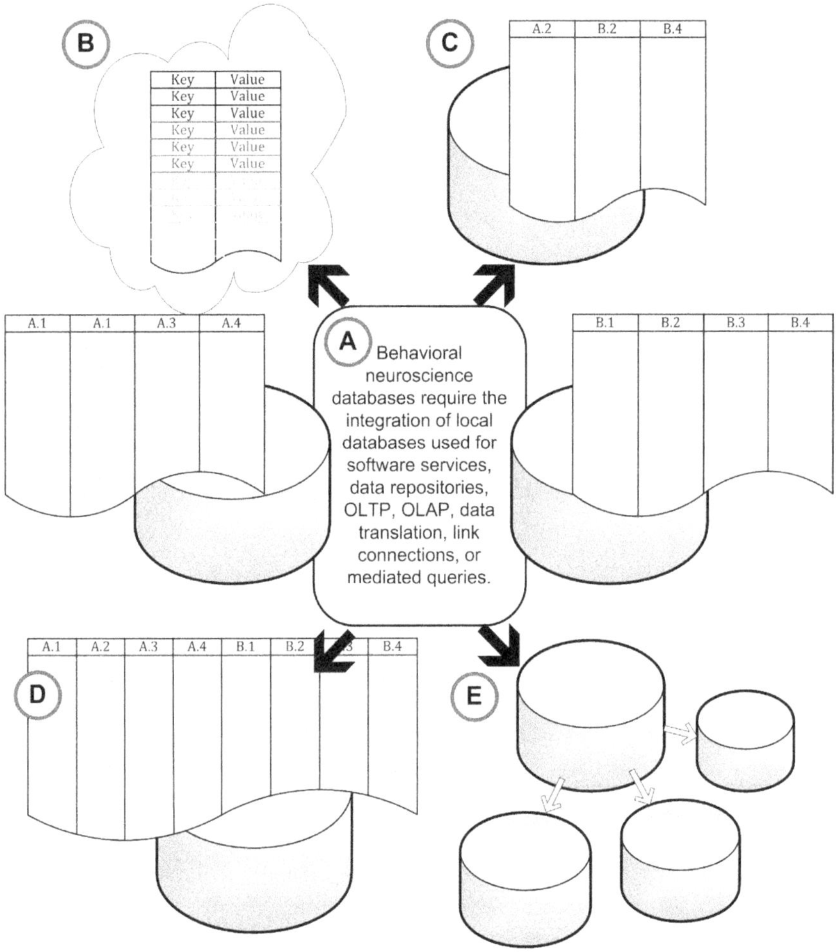

Figure 2.1 Databases interact with nearly all aspects of biological science. The ubiquitous and transparent nature of relational databases places them near the center of numerous bioinformatics functions in neuroscience. (A) They serve as local and community data repositories, the backend for numerous software services, and data sources for translating information between domains. Convergence of relational databases may be through (B) non-strict NoSQL databases, (C) federated databases, or (C) data warehouses. (D) Each approach can use either local or distributed database architectures.

and mass spec data (Horai et al., 2010), behavioral data (Maddatu, Grubb, Bult, & Bogue, 2012), electrophysiological measurements (Günay et al., 2009), and a series of disorder related repositories (Goodman et al., 2003; Matuszek & Talebizadeh, 2009). While not necessarily in conflict with the strategic goals of the greater behavioral neuroscience community, the *ad hoc* collection of boutique databases, analysis tools and information repositories that exist on the local level are often incompatible with comprehensive data mining. This incompatibility arises from an inability to accurately communicate and translate between individual repositories and the lack of a globally definable workflow that can be used to shape a universal strategy.

Even within behavioral neuroscience, multiple data mining strategies exist to identify the causative molecular profile of a given disease model, leading the community to recognize the need to maximize data mining flexibility across all information sources in order to support the iterative hypothesis generation, testing, and observation cycle implicit in the scientific method of life science. The goal of rapidly identifying putative and testable hypotheses about genes or proteins as they relate to behavioral neuroscience disorders has shaped the way next-generation bioinformatics databases integrate data across domains. Some, such as the NeuroCommons Project, attempt to create open-source knowledge frameworks that can integrate diverse data sets at the level of semantics and natural language processing (Ruttenberg et al., 2009). Others, such as GeneWeaver (Baker, Jay, Bubier, Langston, & Chesler, 2012) and GeneNetwork (Wang, Williams, & Manly, 2003), rely almost wholly on the semi-automated integration of primary and secondary data across broad genomics or genetics data sets. Still others, like the Neuroscience Information Framework (NIF; see Chapter 3), attempt to federate data and information across an entire range of databases and independent data sets (Gardner et al., 2008).

Regardless of which strategic approach to database integration the behavioral neuroscience community converges upon, individual researchers or collaborations at the local level should be focused on keeping data in a self-consistent structured and annotated format. Databases, with their ubiquitous presentation, provide the best option for the broadest range of data structures. While numerous strategies exist to integrate databases at several levels, a minimal understanding of how databases function can help guide the discussion of these infrastructure options. More importantly, the landscape of databases available to novice and expert users continues to grow, providing numerous new options for managed data access and integration of intra- and interdisciplinary data.

3. DATABASES: UNDER THE HOOD

3.1. A generalized solution

A database can be generalized to include any intentional system used to structure data for purposes of storage or retrieval. While all sorts of trivial items fit this definition, including phone books, excel spreadsheets, and this very publication, the idealized notion of a database is often thought of as the ubiquitous electronic repository providing data support for specific domains. The primary discerning difference between the former and latter examples are that the latter has a programmatically managed software layer that interacts with the underlying data and data structure, optimizing both the physical and virtual placement of data to expedite data retrieval, increase fault tolerance, and minimize data redundancy. The overarching management layer, or DBMS, uses, in one way or another, a well-indexed snapshot of its managed data to direct the search, retrieval, import, and annotation of stored information. In practical terms, a successful database would enhance data portability, compatibility (translation), extensibility (ease of annotation and curation), and, importantly, data interoperability and querying.

The concept of a database is deeply and correctly coupled with the concept of data querying. This concept should be familiar to cognitive scientists who consider the processes of memory storage and retrieval. For example, databases are analogous to memory recall (retrieval without cues) and recollection (memory reconstruction) but require sophisticated DBMS systems and structured schemas to optimize the query models. More complex types of memory retrieval, such as recognition or relearning, might be loosely synonymous to concepts of data browsing and data mining, respectively, where complex patterns can be dynamically detected and internalized for future reference. Unlike organisms, however, mechanistic approaches to these advanced data recovery processes require highly efficient data organizing structures and are tightly coupled to procedural algorithms. The analogy of behavioral neuroscience, in general, to information technology is often locally correct but globally insufficient. Fundamentally, for example, living organisms perform better and more efficiently on increasingly complex tasks while information technology becomes increasingly slow and hopelessly deficient as task complexity increases. Young children can manage the intricate semantics of language but have a difficult time multiplying four digit numbers together; computers are optimized to solve the inverse set of problems (Von Foerster, 1967). The limitation of databases, in many ways, is our expectation of precise calculations given the fuzzy inconsistencies of data.

3.2. The database explosion

The explosion of biological database adoption among researchers, many in laboratories without dedicated informatics infrastructures, is driven in large part by need as the types and scope of data produced by modern technologies far outpaces our ability to properly collate the data. To illustrate this point, the Human Genome alone would occupy over 180,000 pages when printed out at a 4.5-point font, and finding meaningful information within it would require equally inefficient volumes of indexed data. Compounding the obviously unmanageable scale of data, there is the need to articulate an endless variety of data types, spanning character-based data, images and proprietary data types. The generic notion of a database is designed explicitly to mediate the centrality of these issues.

The drastic increase in database requirements coincided with the emergence of sophisticated open-source relational DBMS, such as MySQL and PostgreSQL. These systems brought free, robust, and flexible relational databases into the realm of the average biologist, effectively removing the need of costly unsupportable informatics overhead associated with proprietary systems such as Oracle or DB2. Biologists, in turn, began to effectively spread boutique bioinformatics databases with minimal entry requirements. The emergence of need and the ubiquitously standardized relational database has pushed researchers to adopt practices that only a decade ago seemed insurmountable. They have embraced a digitized life; gained an appreciation, albeit a subconscious one, of atomic data types, have rationalized the benefits of extensible data models; and have structured future experimentation planning around compatibility.

3.3. Relational databases

The most common incarnation of a DBMS is based on a relational structure. This can be referred to as a Relational DBMS, or RDBMS, where data are structured according to rows and columns. The most common metaphor for visualizing this type of data structure is the spreadsheet, where rapid look-ups are performed by identifying data at the intersection of rows and columns of interest (Fig. 2.2). In both RDBMS and spreadsheets, there is a requirement that data types must be atomic, meaning that they must have a finite scope of values interpretable by computation systems. Any given spreadsheet cell must be either referenced as a number or character, not as both. In many nonbiological databases adherence to atomic data types is easily achieved. This is not necessarily the case with biological data, which can often be described as fuzzy, making it difficult to find items that have continuous similarity with other items. For example, the spectrum of observable phenotypes, characterized by complex disorders like autism, alcoholism, or drug addiction, do not by themselves reference the full spectrum of underlying functional processes motivating their presentation. As a result, the vast

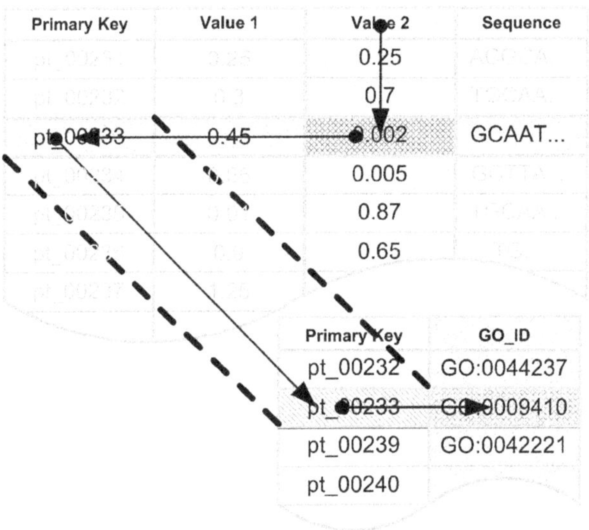

Figure 2.2 The semantic of a relational database. Relational databases rely on strict schemas and data types layered two-dimensional metaphors, where data can be found at the intersection of rows and columns of interest. Strict schematic rules and the use of primary keys ensure a minimization of data redundancy and provides for a mathematically based approach to data querying (SQL).

majority of continuous biological data needs to be extracted from bioinformatics databases and manipulated by independent algorithms.

Finding synergy between diverse data types is often overcome through the creation of elaborate data schemas that attempt to either gather a wide range of very granular data to produce strict data types, or manage only very high-level metadata connections, effectively eliminating the internal database optimizations that are at the core of modern database robustness. In behavioral neuroscience, this is analogous to the pros and cons of losing information within a subset of molecular functions versus losing information about the relationship between the biological processes occupied by those molecular functions.

One major distinction between flat-file data representations, like row by column spreadsheets and NoSQL (Not-only SQL) databases, and RDBMS is that data in relational database schemas are built around a unique identifier for each record, called a *primary key*. A primary key ensures that one and only one instance of an entity or relationship exists and allows database schemas to be optimized to reduce redundancy and query time through a process called normalization. Interestingly, this powerful aspect of a relational database can often serve to complicate their application in biological domains. For example, the word *hypothalamus* can be used as an implicit organizing metaphor for objects relating to stress response, diurnal cycles, metabolism, and thermoregulation, among others, but it does not *uniquely* reference any given atomic (non-divisible) object. Unfortunately, the application of semantic terms, such as "hypothalamus", is wholly ineffective in life science because of the plasticity of language and redundancy of function in biology. While ontologies are useful to relate shared relationships based on collaborative annotations and can substitute, at times, as contextual primary keys, they do not wholly replace the normative database definition of a primary key. In fact, from a strict database perspective, there is a noticeable lack of primary keys in biology, as there exists no emergent or organic descriptor that can reference every known and unknown biological object in perpetuity. As a result, many existing behavioral neuroscience databases use as their reference points objects that may change over time or between contexts. The alcohol-related gene CREB, for example, has references to 77 unique accession numbers in NCBI-related databases, making it nearly impossible to pinpoint a canonical definition.

Another consequence of the structure imposed by RDBMS is the creation of a standardized declarative query language. Based on mathematical concepts of relational algebra and tuple relational calculus, SQL (Structured Query Language) provides set logical and procedural ways to interact with data in a context that is independent of the relational database vendor (see Berenson et al., 1995 for a review). While modern RDBMs shoulder much of the burden for query optimization and load balancing, the concepts driving relational databases are formative to understanding the numerous database variants employed to overcome shortcomings in this approach. Ultimately, the choice of an underlying biological database is a trade off between costs, speed, redundancy, and complexity, all driven by the types of data to be stored.

3.4. Analytical databases

Analytical databases are typically read-only databases that are specifically designed to support data mining on an underlying, mostly static, set of information. They are not designed solely to distribute or house data. Community data repositories that fall into this category are the result of efforts to bring both data and tools that operate on that data under the same information structure. Researchers in behavioral neuroscience interested in sharing a stable set of data while providing interactive tools for integrating primary or secondary data to create new knowledge may gravitate toward these types of resources. Examples in behavioral neuroscience include the Comparative Toxicogenomics Database (Davis, Murphy, Rosenstein, Wiegers, & Mattingly, 2008), MuTrack (Baker et al., 2004), GeneWeaver (Baker et al., 2012), or NCBI's GEO and CDART (Sayers et al., 2012). As information processing becomes more seamlessly integrated with database infrastructures there is a trend to include analytics at the user interface level, but this trend is limited by the complexity of the analytics and the scope of the information to be mined. Dynamic analytics at the user interface level, for example, do not perform well in complex (or genome-scale) tasks that require prolonged periods of time to accomplish. Advances in high-performance computing algorithms are mitigating this challenge (Chesler & Langston, 2006).

3.5. Data warehouse

Data warehouses are effective for behavioral scientists desiring to integrate and distribute data without embedding an analytics framework (Keator, 2009). As the name indicates, data warehouses are explicitly designed to store data under a common framework. Individual operation systems, located locally or disparately, contribute information through a shared integration layer to a central repository. Through this process of integration, data is *cleansed*, or transformed to meet homogeneous criteria. Unfortunately, the process of data cleansing often leads to lossy data constructs, where the original data may not be recapitulated. On the other hand, centralized data repositories can easily be subdivided into functional domains of interest, referred to as "data marts," like BioMart (Haider et al., 2009). In neuroscience, data warehouses are manifested in several efforts to collect and unify data under consistent schemas. There are domain-specific data centers, such as BrainMap (www.brainmap.org), which stores functional neuroimaging literature, and PubBrain (www.pubbrain.org), which communicates directly with the PubMed data warehouse, and broader community efforts. The NIF is an example of a community data warehouse that contains a registry of over 4800 individual data or metadata resources (Gardner et al., 2008).

3.6. Federated databases

Federated databases were originally described as a set of autonomous databases that promote unified access through a set of structured meta-data fields (see Heimbigner & McLeod, 1985). This approach has been more loosely applied to include *composite* databases, which are transparent integrations of autonomous database systems under a globally mandated schema. In both cases, integration is done at the level of common meta-data architecture. Federated databases can be either locally centralized or geographically distributed, and occupy a level autonomy that ranges from loosely coupled to tightly coupled federated schemas. Good examples in behavioral neuroscience include NIF (Gardner et al., 2008) and the Biomedical Informatics Research Network (Ashish, Ambite, Muslea, & Turner, 2010). While the vast majority of behavioral neuroscience laboratories lack the technical skills to navigate the implementation of their own federated databases, they can mediate the exchange of their data with these robust repositories by intentional efforts of data standardization. Minimal Information Standards can be used to provide a common framework to integrate data. Minimum Information for Biological and Biomedical Investigations (Taylor et al., 2008) or Minimal Information About Neural Electromagnetic Ontologies (Frishkoff et al., 2011; see also Chapter 15) are two examples.

3.7. Laboratory information management systems

The most prevalent type of data resource within behavioral neuroscience is the LIMS. These predominantly local systems are developed over time to meet the specific needs of a given laboratory or research group and are often not designed *de novo* to integrate data with external resources. In many cases, several LIMS coexist to capture varying parts of the information landscape. Wikis, for example, provide an excellent means for capturing the free-form concepts of an electronic laboratory notebook, where students and investigators can collaborate and develop institutional memory about protocols and experimental results (Waldrop, 2008). Larger collaborations may choose highly structured to LIMS to track samples and provide a layer of analytics (Baker et al., 2004). These types of LIMS systems often require dedicated informatics objectives and resources but can be built upon readily available technologies. While no single resource exists to satisfy the LIMS needs of every situation, domain-specific LIMS can address the management of particular technologies. The BioArray Software Environment is designed to manage microarray data (Saal et al., 2002), while the BioGRID is a general purpose repository for interaction datasets (Stark et al., 2006). Commercial solutions exist, as well, but they can limit researchers into a proprietary framework than does not necessarily promote flexibility.

3.8. Knowledge bases

Many consortium projects, programs, model organism communities, and collaborative efforts bring together widely diverse research approaches and resources around a particular area of investigation. These specialized databases are designed to logically represent information repositories to aid in decision-making processes and can include white papers, FAQs, user manuals, tutorials, encyclopedias, dictionaries, and other forms of flat files. Wiki-omics (Waldrop, 2008), in neuroscience, for example, provides a good example for this type of free-form data organized around intuitive or pre-identified relationships. Machine-readable databases attempt to make logical connections between data and data types by relying on the semi-structured annotation of the underlying data. Ontologies in neuroscience can leveraged for annotation of unstructured data. The NCBO annotator, for example, can be used to automate the context of free-form data by attaching semantic meaning to ontological frameworks (Jonquet, Shah, & Musen, 2009). Similarly, the NIF have leveraged Texpresso for similar purposes to locate and extract data from the literature (Bandrowski et al., 2012; Müller et al., 2008). Machine and human-driven knowledge bases can therefore be successfully combined to navigate data using both approaches.

4. BEYOND RELATIONAL DATABASES

As the scope and depth of data within behavioral neuroscience databases rapidly expands, the commensurate increase in relational database complexity and size consequently limits retrieval times, restricts exhaustive integration, and requires increasingly more overhead and expertise to manage. Since early 2009, there has been an intentional effort to circumvent these complexity drawbacks by implementing a type of database referred to as NoSQL databases. These databases, while not technically relational databases since they lack traditional mechanisms that would allow for normalization, have the benefit of being natively optimized for popular cloud-based and multicore computer architectures. They are designed to discover data in extremely large data sets at speeds that rival and surpass the performance of large parallel databases without many of the drawbacks (Stonebraker et al., 2010). Since NoSQL databases lack traditional schemas, there are few limiting requirements for time-consuming database administration and can be managed through low-level application programming interfaces instead of optimized SQL queries.

4.1. Wide column and key-value stores

The removal of tightly controlled data schemas, which effectively denormalizes data structures and therefore greatly increases the risk of redundancy, is compensated for by creating operations that are (1) easily deployed and (2) natively distributed. Hadoop (Shvachko, Kuang, Radia, & Chansler, 2010), an open-source implementation of MapReduce (Dean & Ghemawat, 2008), is an example of a key-value long table. Similar to Google's BigTable implementation (Chang et al., 2008), Hadoop relies exclusively on the qualities of well-indexed data to very rapidly discover values associated with particular keys, called key-value pairs. When implemented properly and for purposes of finding one-to-one or one-to-many associations with a key of interest, Hadoop delivers the power of large and expensive parallel RDBMS without any of the overhead. Other popular implementations of MapReduce include Cassandra and Amazon's SimpleDB. While they may perform extremely well in data location and retrieval, they sub-perform under a range of scenarios, including determining data consistency and transaction control, which are pushed back to the user or the interface controller. Regardless, the future of these types of data structures is very bright in areas of biological databases where querying specific entities within voluminous data stores is a common task.

4.2. Document stores

The contemporary version of the flat-file database is referred to as a document-oriented NoSQL database, sometimes known as the document store. Here, databases such as MongoDB, CouchDB, and OrientDB, among others, are optimized specifically for indexed JSON-styled documents (Banker, 2012; Wei, Sicong, Qian, & Amiri, 2009). They form the backbone of many web services required to rapidly distribute large numbers of records, including increasingly popular web streaming content. While not used in any current large-scale behavioral neuroscience effort, the document store's reliance on NoSQL's key-value relationship schema places it in the unique position of being able to satisfy growing data needs without costly infrastructure support. Indeed, schemas in document stores are dynamically generated and can scale to meet nearly all data types.

4.3. Graph databases

Systems biology, largely centered on the analysis of biological networks, is becoming increasingly widely applied in neuroscience. There exists no shortage of topological life science domains that currently incorporate networks (and therefore the underlying graph theory) for the elucidation of specific processes. Behavioral neuroscience, for example, is interested in the descriptive and predictive potentials of how the underlying gene, protein or metabolic network relationships effect complex traits (Spanagel, 2009). Of paramount importance is the discovery of unifying principles mediating network topology and their biological relevance. There is a need to understand how large-scale interacting dynamical systems, such as those found in systems biology, behave collectively (Strogatz, 2001); empirical studies have shed light on the topology of cellular and metabolic networks (Bhalla & Iyengar, 1999; Hartwell, Hopfield, Leibler, & Murray, 1999; Veeramani & Bader, 2010) and neural networks (Kim, 2004). The extension of graph theory into the collective analysis of behavioral neuroscience networks provides a tremendous reservoir of qualitative insight into the function of biological systems under equilibrium and dynamic stresses.

This has led to an urgent need to refine computational models for graph pattern mining and a robust means for storing, collating, and translating across immense genome-scale graphs in a way that supports the global application of appropriate analysis tools. Because there exists no relational database model applicable across large heterogeneous data representations (and, consequently, repositories) of graph/network-based approaches to biological data, several NoSQL models have made rapid progress to close the gap. These approaches use key-value relationships to generalize pairwise and tripartite relationships between unbounded numbers of biological data types, creating general graph-based schemas that are optimized for generically applied networks and semantic web information. These include Neo4j (and its biology relative, Bio4j), AllegroGraph, sones, infogrid, and trinity, among others. Other graph-based efforts are focusing on compatible labeled graph formats represented by the web-based RDF schemas (Belleau, Nolin, Tourigny, Rigault, & Morissette, 2008; Mironov et al., 2012). The NIF and semantic enterprise wiki from the Allen Institute rely, in part, on graph databases.

5. LIVING WITH HETEROGENEITY

5.1. Integrating primary data

The hierarchical complexities and layered dependencies underlying the continuum of observable processes in behavioral neuroscience result in an inability of a single researcher to encapsulate an effective scope of knowledge. Perhaps the paramount success of bioinformatics is the recognition that deep understanding is found at the intersection of multiple data domains and data types across physiological, developmental, and evolutionary time scales. This can be done by articulating primary data across numerous domains and has led to several emergent realities: (1) structured vocabularies and experimental protocols provide a foundational framework designed to enhance integration, (2) federated databases operate more efficiently on highly structured data, and (3) data needs to be valued as open-source resources (Chesler & Baker, 2010).

Structured vocabularies and ontologies are well-defined controlled vocabularies designed to formalize interactions within the broad scope of experimental observations. However, for each approach to structured integration, there is a tradeoff between prescription and flexibility. As data attributes become more highly structured, the underlying database becomes more accurate and efficient, but at the same time more narrowly defined. In life science, this is the tension between a narrow scope that returns false negatives and articulations that are too broadly defined to be informative. Compounding this tension is a competing tradeoff between the often labor-intensive process required to hand-curate narrowly defined domains and the computationally efficiency associated with automated or semi-automated data management. These manifest themselves in the type of connections established between data sets, from low-level link connections (SRS; Etzold, Ulyanov, & Argos, 1996) and mediated queries (TAMBIS (Stevens et al., 2000) and Kleisli (Davidson et al., 2001)) to full integration. For example, domain-specific and generalized ontologies, such as NIF's NeuroLEX or GO (Ashburner et al., 2000), respectively, are intended to provide translational flexibility at the interface of databases and analysis tools and are excellent pivot objects in mediated data sources. However, ontologies are not error free and may be considered too sparse or biased to cover an appropriate range of represented system states in a completely automated fashion. The significant challenges in the construction of an ontology that spans all behavioral neuroscience is representative of this problem.

One interesting core aspect of RDBMS is their definitional use of primary keys for the purposes of normalization and uniquely defining relationships of interest, ideally allowing for the harmonization of data between data sources. A primary key uniquely defines an object and remains temporally and contextually constant. Life science is unique in that there exists no global organic primary key. While genes are often used as a core organizing metaphor, they do not have the benefit of remaining contextually constant. The concept that even trusted biological objects shift in both meaning and value over time is a well-known and primary distinction between biological databases and other enterprise level databases (Birney & Clamp, 2004). Thus, primary data is often organized around relative relationships between objects or data types of interest. Automating the discovery of relative relationships between databases is a difficult task that requires the constant curation of information, even in federated environments where strict rules are applied, and often relies heavily on ontological relationships. NoSQL data stores have the benefit of not having to contend with primary keys or strict schemas, lowering the difficulty of dealing with shifting definitions of reference sources.

5.2. Managing secondary data

One approach to reduce confounding background clatter of data with low information content is to focus database integration efforts on published or peer-reviewed data sets. Since these data sets are often representative and significant subsets of larger primary data pools, they are referred to as secondary data. In many ways, the neuroscience bioinformatics is a leader in this area, with efforts like the Neuroinformatics Framework (NIF) (Gardner et al., 2008) and GeneWeaver (Baker et al., 2012), where data stores are integrated at the most granular level of discrete object relationships.

As efforts to collect and collate neuroscience data have discovered, there is a clear imperative to scraping secondary data from published material. Printed academic journals have been slow to standardize the format of primary and supplemental content. For example, while most journals accept Microsoft-based publication standards, reading in data from a table requires both the digitized access to the document and a curator to determine the context of the information. One strong argument for the tacit use of ontologies and structured vocabularies is to further enforce a machine-readable context for published secondary data to the extent that biological databases will eventually merge with journals to seamlessly integrate data. The use of Uniform Resource Identifiers for uniquely referencing particular entities will further such capabilities. Capturing digitized primary and secondary data in a NoSQL-Journal hybrid approach, for example, also allows for the capture of data provenance. While there is a high-level practical need to track data, there is a cultural need to indicate data generation and sourcing in order to encourage researchers to share, and ultimately enhance, knowledge production and aggregation.

Another interesting phenomenon of secondary data analysis is that data aggregation over these sets indicates a strong asymmetry in data density. This means that observable associations between certain biological objects are consistent over a wide range of data sets. This observation, known as a scale-free network in graph theory (Wolf, Karev, & Koonin, 2002), is a well-recognized phenomenon of primary data interactions in biological data, but was unfamiliar in broad secondary or federated data sets. The observation of data sparsity over data scarcity has implications in how neuroscience databases should think about internal schemas. For example, if a database is tasked with storing data about molecular networks in behavioral neuroscience and discovering information about the shortest path between objects of interest, then storing data in an edge list is much better for handling algorithms associated with shortest path problems in sparse networks. It also indicates that in the practical and esoteric world of database, the volume of data does not always relate to information or importance of that data.

6. CONCLUSION

Bioinformatics is fundamentally about the information of biology. Information, in turn, is buried within a cacophony of data produced by a wide swath of molecular techniques. In neuroscience, the breadth of data is exceptionally large as it spans genomics, proteomics, metabolomics, image analysis, and behavioral science, among other protocols, and requires researchers to store data with due diligence based on the data types, data scope and depth, and underlying querying requirements. Traditional relational databases can effectively manage data but require in-depth domain knowledge and strong database expertise to produce schemas robust enough to handle scope and integration. The emergence of NoSQL databases in the recent years has caused researchers to reexamine how data is structured and explore flexible alternatives for viewing relationships among differing data types typically encountered in behavioral neuroscience.

REFERENCES

Altman, R. B. (2004). Building successful biological databases. *Briefings in Bioinformatics*, *5*, 4–5.

Ashburner, M., Ball, C. A., Blake, J. A., Botstein, D., Butler, H., Cherry, J. M., et al. (2000). Gene ontology: Tool for the unification of biology. The Gene Ontology Consortium. *Nature Genetics*, *25*, 25–29.

Ashish, N., Ambite, J. L., Muslea, M., & Turner, J. A. (2010). Neuroscience data integration through mediation: an (F)BIRN case Study. *Frontiers in Neuroinformatics*, *4*, 118.

Baker, E. J., Galloway, L., Jackson, B., Schmoyer, D., & Snoddy, J. (2004). MuTrack: A genome analysis system for large-scale mutagenesis in the mouse. *BMC Bioinformatics*, *5*, 11.

Baker, E. J., Jay, J. J., Bubier, J. A., Langston, M. A., & Chesler, E. J. (2012). GeneWeaver: A web-based system for integrative functional genomics. *Nucleic Acids Research, 40*, D1067–D1076.

Bandrowski, A. E., Cachat, J., Li, Y., Muller, H. M., Sternberg, P. W., Ciccarese, P., et al. (2012). A hybrid human and machine resource curation pipeline for the Neuroscience Information Framework. *Database, 2012*, bas005.

Banker, K. (2012). *MongoDB in Action*. Shelter Island, NY: Manning.

Belleau, F., Nolin, M.-A., Tourigny, N., Rigault, P., & Morissette, J. (2008). Bio2RDF: Towards a mashup to build bioinformatics knowledge systems. *Journal of Biomedical Informatics, 41*, 706–716.

Berenson, H., Bernstein, P., Gray, J., Melton, J., O'Neil, E., & O'Neil, P. (1995). *A Critique of ANSI SQL Isolation Levels*. ACM Press pp. 1–10.

Bhalla, U. S., & Iyengar, R. (1999). Emergent properties of networks of biological signaling pathways. *Science, 283*, 381–387.

Birney, E., & Clamp, M. (2004). Biological database design and implementation. *Briefings in Bioinformatics, 5*, 31–38.

Chang, F., Dean, J., Ghemawat, S., Hsieh, W. C., Wallach, D. A., Burrows, M., et al. (2008). Bigtable. *ACM Transactions on Computer Systems, 26*, 1–26.

Chesler, E. J., & Baker, E. J. (2010). The importance of open-source integrative genomics to drug discovery. *Current Opinion in Drug Discovery & Development, 13*, 310–316.

Chesler, E., & Langston, M. (2006). Combinatorial genetic regulatory network analysis tools for high throughput transcriptomic data. In E. Eskin, T. Ideker, B. Raphael & C. Workman (Eds.), *Systems Biology and Regulatory Genomics* (pp. 150–165). Berlin/Heidelberg: Springer.

Colland, F., Jacq, X., Trouplin, V., Mougin, C., Groizeleau, C., Hamburger, A., et al. (2004). Functional proteomics mapping of a human signaling pathway. *Genome Research, 14*, 1324–1332.

Davidson, S. B., Crabtree, J., Brunk, B. P., Schug, J., Tannen, V., Overton, G. C., et al. (2001). K2/Kleisli and GUS: Experiments in integrated access to genomic data sources. *IBM Systems Journal, 40*, 512–531.

Davis, A. P., Murphy, C. G., Rosenstein, M. C., Wiegers, T. C., & Mattingly, C. J. (2008). The Comparative Toxicogenomics Database facilitates identification and understanding of chemical-gene-disease associations: Arsenic as a case study. *BMC Medical Genomics, 1*, 48.

Dean, J., & Ghemawat, S. (2008). MapReduce. *Communications of the ACM, 51*, 107.

Etzold, T., Ulyanov, A., & Argos, P. (1996). SRS: Information retrieval system for molecular biology data banks. *Methods in Enzymology (Elsevier), 266*, 114–128.

Frishkoff, G., Sydes, J., Mueller, K., Frank, R., Curran, T., Connolly, J., et al. (2011). Minimal Information for Neural Electromagnetic Ontologies (MINEMO): A standards-compliant method for analysis and integration of event-related potentials (ERP) data. *Standards in Genomic Sciences, 5*, 211–223.

Gardner, D., Akil, H., Ascoli, G. A., Bowden, D. M., Bug, W., Donohue, D. E., et al. (2008). The neuroscience information framework: a data and knowledge environment for neuroscience. *Neuroinformatics, 6*, 149–160.

Goodman, N., McCormick, K., Goldowitz, D., Hockly, E., Johnson, C., Kristal, B., et al. (2003). Plans for HDBase—A research community website for Huntington's Disease. *Clinical Neuroscience Research, 3*, 197–217.

Günay, C., Edgerton, J. R., Li, S., Sangrey, T., Prinz, A. A., & Jaeger, D. (2009). Database analysis of simulated and recorded electrophysiological datasets with PANDORA's toolbox. *Neuroinformatics, 7*, 93–111.

Haider, S., Ballester, B., Smedley, D., Zhang, J., Rice, P., & Kasprzyk, A. (2009). BioMart Central Portal—Unified access to biological data. *Nucleic Acids Research, 37*, W23–W27.

Hartwell, L. H., Hopfield, J. J., Leibler, S., & Murray, A. W. (1999). From molecular to modular cell biology. *Nature*, *402*, C47–C52.

Heimbigner, D., & McLeod, D. (1985). A federated architecture for information management. *ACM Transactions on Information Systems*, *3*, 253–278.

Horai, H., Arita, M., Kanaya, S., Nihei, Y., Ikeda, T., Suwa, K., et al. (2010). MassBank: A public repository for sharing mass spectral data for life sciences. *Journal of Mass Spectrometry*, *45*, 703–714.

Jonquet, C., Shah, N. H., & Musen, M. A. (2009). The open biomedical annotator. *Summit on Translatation Bioinformatics*, *2009*, 56–60.

Keator, D. B. (2009). Management of information in distributed biomedical collaboratories. *Methods in Molecular Biology*, *569*, 1–23.

Kim, B. J. (2004). Performance of networks of artificial neurons: The role of clustering. *Physical Review. E, Statistical, Nonlinear, and Soft Matter Physics*, *69*, 045101.

Maddatu, T. P., Grubb, S. C., Bult, C. J., & Bogue, M. A. (2012). Mouse Phenome Database (MPD). *Nucleic Acids Research*, *40*, D887–D894.

Marcus, D. S., Wang, T. H., Parker, J., Csernansky, J. G., Morris, J. C., & Buckner, R. L. (2007). Open Access Series of Imaging Studies (OASIS): Cross-sectional MRI data in young, middle aged, nondemented, and demented older adults. *Journal of Cognitive Neuroscience*, *19*, 1498–1507.

Matuszek, G., & Talebizadeh, Z. (2009). Autism Genetic Database (AGD): A comprehensive database including autism susceptibility gene-CNVs integrated with known noncoding RNAs and fragile sites. *BMC Medical Genetics*, *10*, 102.

Mironov, V., Seethappan, N., Blondé, W., Antezana, E., Splendiani, A., & Kuiper, M. (2012). Gauging triple stores with actual biological data. *BMC Bioinformatics*, *13* (Suppl. 1), S3.

Müller, H.-M., Rangarajan, A., Teal, T. K., & Sternberg, P. W. (2008). Textpresso for neuroscience: Searching the full text of thousands of neuroscience research papers. *Neuroinformatics*, *6*, 195–204.

Ruttenberg, A., Rees, J. A., Samwald, M., & Marshall, M. S. (2009). Life sciences on the Semantic Web: The Neurocommons and beyond. *Briefings in Bioinformatics*, *10*, 193–204.

Saal, L. H., Troein, C., Vallon-Christersson, J., Gruvberger, S., Borg, A., & Peterson, C. (2002). BioArray Software Environment (BASE): A platform for comprehensive management and analysis of microarray data. *Genome Biology*, *3*, SOFTWARE0003.

Sayers, E. W., Barrett, T., Benson, D. A., Bolton, E., Bryant, S. H., Canese, K., et al. (2012). Database resources of the National Center for Biotechnology Information. *Nucleic Acids Research*, *40*, D13–D25.

Shepherd, G. M., Mirsky, J. S., Healy, M. D., Singer, M. S., Skoufos, E., Hines, M. S., et al. (1998). The Human Brain Project: Neuroinformatics tools for integrating, searching and modeling multidisciplinary neuroscience data. *Trends in Neurosciences*, *21*, 460–468.

Shoemaker, B. A., Zhang, D., Tyagi, M., Thangudu, R. R., Fong, J. H., Marchler-Bauer, A., et al. (2012). IBIS (Inferred Biomolecular Interaction Server) reports, predicts and integrates multiple types of conserved interactions for proteins. *Nucleic Acids Research*, *40*, D834–D840.

Shvachko, K., Kuang, H., Radia, S., & Chansler, R. (2010). The Hadoop Distributed File System. IEEE 26th Symposium On Mass Storage Systems and Technologies (MSST), pp. 1–10.

Spanagel, R. (2009). Alcoholism: A systems approach from molecular physiology to addictive behavior. *Physiological Reviews*, *89*, 649–705.

Stark, C., Breitkreutz, B.-J., Reguly, T., Boucher, L., Breitkreutz, A., & Tyers, M. (2006). BioGRID: A general repository for interaction datasets. *Nucleic Acids Research*, *34*, D535–D539.

Stevens, R., Baker, P., Bechhofer, S., Ng, G., Jacoby, A., Paton, N. W., et al. (2000). TAMBIS: Transparent access to multiple bioinformatics information sources. *Bioinformatics*, *16*, 184–186.

Stonebraker, M., Abadi, D., DeWitt, D. J., Madden, S., Paulson, E., Pavlo, A., et al. (2010). MapReduce and parallel DBMSs: Friends or foes? *Communications of the ACM*, *53*, 64–71.

Strogatz, S. H. (2001). Exploring complex networks. *Nature*, *410*, 268–276.

Taylor, C. F., Field, D., Sansone, S.-A., Aerts, J., Apweiler, R., Ashburner, M., et al. (2008). Promoting coherent minimum reporting guidelines for biological and biomedical investigations: The MIBBI project. *Nature Biotechnology*, *26*, 889–896.

Veeramani, B., & Bader, J. S. (2010). Predicting functional associations from metabolism using bi-partite network algorithms. *BMC Systems Biology*, *4*, 95.

Von Foerster, H. (1967). Biological principles of information storage and retrieval. In A. Kent, O. E. Taubee, J. Beltzer & G. D. Goldstein (Eds.), *Electronic Handling of Information: Testing and Evaluation* (pp. 123–147). London: Academic Press.

Waldrop, M. (2008). Big data: Wikiomics. *Nature*, *455*, 22–25.

Wang, J., Williams, R. W., & Manly, K. F. (2003). WebQTL: Web-based complex trait analysis. *Neuroinformatics*, *1*, 299–308.

Wei, K., Sicong, T., Qian, X., & Amiri, H. (2009). An Investigation of No-SQL Data Stores. Most.

Wolf, Y. I., Karev, G., & Koonin, E. V. (2002). Scale-free networks in biology: New insights into the fundamentals of evolution? *BioEssays*, *24*, 105–109.

CHAPTER THREE

A Survey of the Neuroscience Resource Landscape: Perspectives from the Neuroscience Information Framework

Jonathan Cachat, Anita Bandrowski, Jeffery S. Grethe, Amarnath Gupta, Vadim Astakhov, Fahim Imam, Stephen D. Larson, Maryann E. Martone[1]

Department of Neurosciences and Center for Research in Biological Systems, University of California, San Diego, California, USA

[1]Corresponding author: e-mail address: mmartone@ucsd.edu

Contents

Abstract

The number of available neuroscience resources (databases, tools, materials, and networks) available via the Web continues to expand, particularly in light of newly implemented data sharing policies required by funding agencies and journals. However, the nature of dense, multifaceted neuroscience data and the design of classic search engine systems make efficient, reliable, and relevant discovery of such resources a significant challenge. This challenge is especially pertinent for online databases, whose dynamic content is largely opaque to contemporary search engines. The Neuroscience Information Framework was initiated to address this problem of finding and utilizing neuroscience-relevant resources. Since its first production release in 2008, NIF has been surveying the resource landscape for the neurosciences, identifying relevant resources and working to make them easily discoverable by the neuroscience community. In this chapter, we provide a survey of the resource landscape for neuroscience: what types of

International Review of Neurobiology, Volume 103
ISSN 0074-7742
http://dx.doi.org/10.1016/B978-0-12-388408-4.00003-4

resources are available, how many there are, what they contain, and most importantly, ways in which these resources can be utilized by the research community to advance neuroscience research.

1. INTRODUCTION

The availability of a significant portion of humanity's knowledge through the World Wide Web is an achievement of momentous significance. Standardization of protocols for posting files, images, and other data objects along with the parallel development of search engines and Web portals for discovering information has potentiated the dawn of a new age in scientific communication (Hey, Stewart, & Kristin, 2004). The central challenge of our time is developing ways to uncover knowledge within the vast amounts of data awaiting comparison, integration, and interpretation (Akil, Martone, & Van Essen, 2011; Kötter, 2001). Scientific data, however, relies on considerable contextual information to make results interpretable (Martone, Gupta, & Ellisman, 2004) and for this reason the development of (semi-) automated scientific knowledge discovery systems is particularly difficult (Barnes and Shaw, 2009). Moreover, beyond the pharmaceutical domain, there is relatively small commercial potential in such informatics mining efforts, suggesting that scientists will have to take it upon themselves to adopt best practices and put forth solutions for facilitating scientific data exchange and knowledge discovery across the Web.

Neuroscience presents a challenging domain for the development of a framework to facilitate data exchange and integration. As an inherently interdisciplinary science, neuroscience provides data from genomic to behavioral levels of analysis, and across ionic to evolutionary temporal scales. From this diversity, researchers focusing at different scales, using different techniques, generate experimental results in multiple formats that are usually unannotated or annotated with custom vocabularies for describing content and metadata. Today, finding and utilizing individual resources requires considerable human effort, particularly when the goal is to compare one set of experimental results to another. Researchers can easily spend hours a day searching for specific pieces of information or browsing the increasingly rich set of available neuroscience-relevant resources. Therefore, the critical task is to organize this data in a meaningful way, such that it will facilitate insights into the structure and function of the nervous system at and across all spatiotemporal levels of analysis. The challenge is to provide tools that allow for systematic, flexible and efficient user-controlled access to the growing multitude of neuroscience data.

The Neuroscience Information Framework (NIF, http://www.-neuinfo.org) project started in 2006 as an initiative of the NIH Blueprint consortium, in recognition of the need to develop a resource description framework and search strategy for locating, accessing, and utilizing resources available for neuroscience research (Gardner et al., 2008a). As defined here, resources include databases, software/Web-based tools, materials, networks, or information that would accelerate the pace of neuroscience research and discovery. Many of these resources were created through significant investment of government funding but remain largely unknown or underutilized by the research community they were created to serve.

The first phase of the NIF, completed in 2008, provided an overview of the number and type of neuroscience-relevant resources currently available and defined a strategy for providing a coherent framework to promote their discovery by the neuroscience research community (Gupta et al., 2008). These efforts resulted in the first version of the NIF Registry, a catalog of neuroscience-relevant resources annotated with a controlled vocabulary covering multiple dimensions (e.g., organism, nervous system level, and resource type). From an initial ~300 entered at the conclusion of phase one of the project, the NIF Registry has swelled to over 4800 resources to date, and continues to grow. Over 2000 of these are databases, ranging in size from 100's to 100's of millions of records. Dynamic databases are considered part of the "deep" or "hidden" Web, in which content is dynamically generated as a function of a query, contained in attachments or other materials that cannot be effectively indexed and searched by traditional search engine systems (Bergman, 2001).

Although many of the databases listed within the registry are general in scope (e.g. genomic databases), there is clear value for the neurosciences in the data they contain. A consideration of the logistics concludes that an individual researcher simply cannot visit and query some 2000 databases separately; a fact compounded by the existence of custom terminologies, query systems and user interfaces which vary from resource to resource. In this report, we provide a survey of the current landscape of neuroscience-relevant resources from the perspective of NIF's mission to enable and improve searching for and integrating information contained within these resources. We also address some of the practical problems we have encountered in the integration of independently developed, diverse, and messy data. With the recent emphasis both inside and outside of academia on "big data," we consider different models of how neuroscience, perhaps the most information rich of all the sciences, can capitalize on these lessons in support of neuroscience discovery.

2. MATERIALS AND METHODS

2.1. Overview of NIF system

The NIF is freely accessed via a Web portal (http://neuinfo.org). The NIF Web portal provides a semantically enhanced search interface in addition to a set of tools and services for identification, registration, ingestion, and curation of data content. NIF is built upon an open-source platform, using the Lucene suite and Solr for indexing of content with custom components developed when necessary (Gupta et al., 2008). The current NIF Portal and advanced search interface is built upon the Google Web Toolkit platform.

In addition to the NIF Web portal, the system can be accessed through a set of Web services. These services permit programmatic access to NIF vocabulary and data services (http://neuinfo.org/developers/index.shtm). Moreover, some of NIF's content, including the NIF Registry and the NeuroLex knowledge base (http://neurolex.org), is made available in RDF via a SPARQL endpoint.

As will be described below, NIF utilizes an expansive ontology, the NIFSTD, as the semantic framework for integration and search of NIF information sources. Ontology services are provided via the OntoQuest server, an OWL compliant relational database (Gupta et al., 2008). NIF is hosted at the University of California, San Diego in association with collaborators at the California Institute of Technology, George Mason University, Yale University Medical College, and Washington University. Additional technical details of NIF's core components are provided in separate reports (e.g., Bandrowski et al., 2012; Gupta et al., 2008; Imam et al., 2012; Marenco et al., 2010).

2.1.1 Content

NIF maintains an accounting of neuroscience-relevant resources in multiple forms to ensure that broad coverage of the resource landscape is provided. A single search at the NIF portal provides simultaneous query across three distinct catalogs of information (Fig. 3.1):

1. *NIF Registry*: A catalog of >4800 resources, organized by resource types (e.g., database, software tool, service resource) and annotated with keywords from the NIF ontology (NIFSTD).

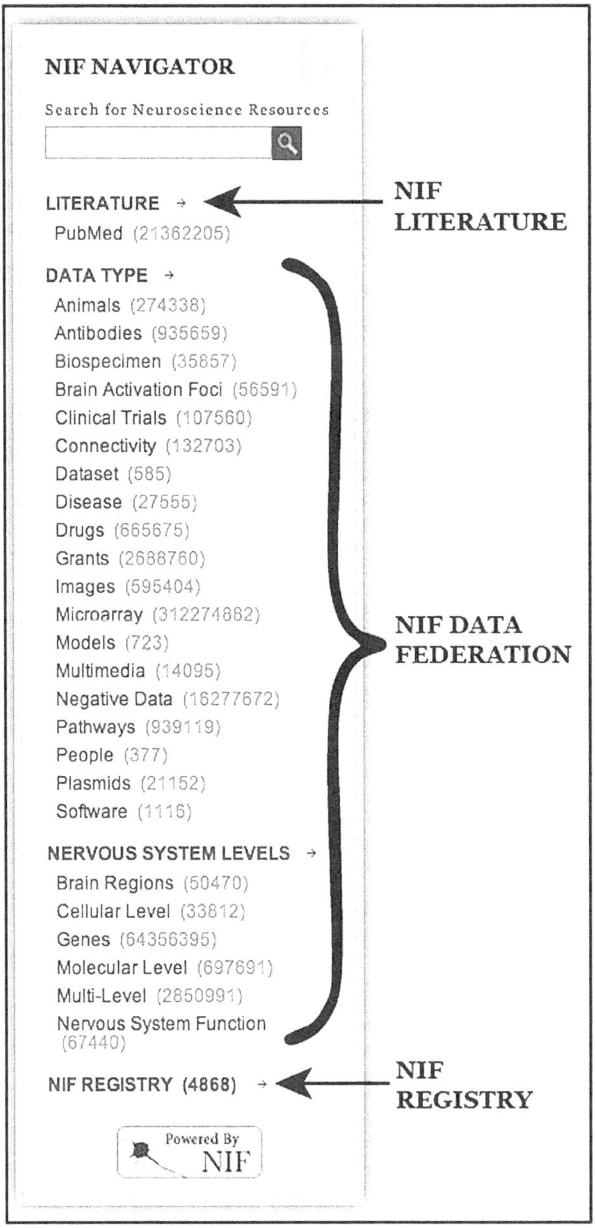

Figure 3.1 NIF Navigator & Overview of NIF Contents. As described, NIF provides simultaneous search over three main indices: (1) NIF Literature, (2) NIF Data Federation, and (3) NIF Registry. The number of records contained in each are shown in gray parentheses following each heading. For the NIF Data Federation, records are organized by Data Type and Nervous System Level, as illustrated in the NIF Navigator. The NIF Navigator is a dynamic, self-contained widget available for download at http://neuinfo.org/downloads/index.shtm. (For color version of this figure, the reader is referred to the online version of this chapter.)

2. *NIF Data Federation*: Deep query into the contents of >150 neuroscience-relevant data databases comprising over 330 million data records, organized by data type and level of the nervous system.
3. *NIF Literature*: Search over 22 million abstracts from PubMed, and full text from open access journals.

NIF allows resources in a variety of formats to be ingested into the data federation (e.g., relational, RDF, XML) via the DISCO tool suite, developed by Marenco, Wang, Shepherd, and Miller (2010). The tool suite includes a centralized dashboard, which allows curators or automated agents to execute scripts that evoke a wide range of functions such as crawling a data source, executing SQL queries, stopping or starting servers, and creating indices. Updates of all resources within the NIF Data Federation are managed by the NIF DISCO scheduler. NIF maintains a full time curator, assisted by several students, responsible for ensuring continual population of the Registry and Data Federation and annotation of NIF content within a consistent annotation framework.

2.1.2 Search

Search is supported by an expansive set of modular ontologies, the NIFSTD (Bug et al., 2008; Imam et al., 2012) covering the main domains of neuroscience. NIFSTD is available via the National Center for Biomedical Ontology's Bioportal (http://bioportal.bioontology.org/ontologies/1084) and also via the NIF Web site (https://confluence.crbs.ucsd.edu/display/NIF/Download+NIF+Ontologies). As a user enters search terms into NIF's Web portal query interface, the system attempts to autocomplete terms from NIFSTD using OntoQuest services. If the search term(s) is contained within NIFSTD, the query is automatically expanded to include synonyms, common abbreviations, and lexical variants. This function represents the semantically enhanced aspects of NIF search and provides a significant advantage of using NIF over other search engines, both general and specific. All of these terms are then joined using an "OR" Boolean operator and treated as one concept. If additional terms are added to the search box, they are joined using "AND" or "OR" operators, depending on the user's selection. The expanded search string used to query NIF content is displayed below the search box and can be edited at will. A "NOT" operator may also be used by manual addition to the search box.

Since 2008, NIF has significantly expanded its concept-based search by including automatic expansion for logically defined classes within the NIFSTD. Defined classes are those classes where membership is inferred via a rule, rather than by direct assertion. OntoQuest flags any defined class in NIFSTD for automatic expansion when that term is selected via autocomplete in the NIF search interface (Imam et al., 2012). For example, NIFSTD contains a list of neurons and a list of small molecules. A module within NIFSTD relates small molecules to neurons through the "has neurotransmitter" property. Thus, a class of neuron can be defined based on its neurotransmitter, for example, a GABAergic neuron is a neuron that uses GABA as a neurotransmitter. When users query for "GABAergic neuron," NIF will automatically expand the search to include all classes of GABAergic neurons currently in NIFSTD based on the "has transmitter property" satisfied with "GABA." NIF also makes extensive use of roles in order to generate useful hierarchies from our existing ontologies. For example, a search for "drug of abuse" will result in a list of small molecules that have the role "drug of abuse." Terms that are defined through their relations are bolded in the autocomplete menu. However, unless a class is defined by an OWL class expression, NIF does not automatically expand the query to include related categories. Rather, the user is given a menu of options through the advanced query interface where they can choose to add related terms as necessary. This strategy was chosen due to the fact that the potential number of related categories can be extremely large (e.g., brain regions). Additionally, this strategy preserves the granularity of a particular query term. For example, if a user searches for a coarse level term like "brain," automatically including any part of the brain may not capture the intent of the query. All NIF vocabulary services are exposed via a set of RESTful Web service calls to Ontoquest so that they can be built into other applications (http://neuinfo.org/developers/index.shtm).

3. RESULTS

The NIF project was created specifically to work with the current state of resources and to provide the capacity for a user to discover relevant resources and utilize their contents more effectively. NIF was not charged with, nor funded for, fielding a unified computational infrastructure for data mining and analytics, although we are beginning to make some tools available for use with NIF's data. Given the state of resources available, NIF designed a practical strategy based on tiers of access to allow maximal exposure of resources, while operating within the fiscal and temporal constraints of both NIF and the resource provider. As the NIF has evolved, the criteria for inclusion within the NIF Registry/Data Federation have changed to adapt to user requests and/or new technologies. In the following sections, we provide an analysis of the current contents of the NIF, based largely on statistics through April 15, 2012. For ease of reference, the URL's for all resources mentioned in the text are included in Table 3.1.

Registry: All resources are registered to the NIF Registry. Resources are identified for inclusion through active outreach by NIF curators, recommendations from the community, and, increasingly, via the NIF automated resource identification pipeline (Bandrowski et al., 2012). The NIF Registry data is hosted by the NeuroLex wiki (http://neurolex.org), a semantic wiki established initially for community maintenance and enhancements of the NIF ontologies (Imam et al., 2012). Each resource receives its own wiki page in NeuroLex, where it can be annotated with NIFSTD terms and keywords.

Currently, the resource Registry is heavily weighted toward databases and software tools (Fig. 3.2A), reflecting NIF's primary purpose and its origin in the Neuroscience Database Gateway, originally developed by the Neuroinformatics Committee of the Society for Neuroscience (NDG; Gardner et al., 2008a). However, over time, we have expanded the Registry to include materials, services, multimedia and training-related resources, based on user requests. NIF also relaxed the policies of the original NDG that excluded genomics and commercial resources, although NIF does not endeavor to have comprehensive or even extensive coverage of commercial products. Indeed, one of the goals of NIF is to promote discovery of NIH-funded resources targeted to the research community, for example, NeuroMab (http://neuromab.org) that may be difficult to find in a Web search without prior knowledge of their existence. Resources must be deemed to be useful for neuroscience researchers, although the resource itself does not have to be neuroscience-focused to meet that criterion. A search for "behavior" in the NIF brings back over 250 data resources.

NIF categorizes the resource based on its type (Fig 3.2) using the NIF resource ontology (http://NeuroLex.org/wiki/Resource_Type_Hierarchy; also available as module within NIFSTD). This ontology was developed independently by the NIF project from initial work by Gardner, Goldberg, Grafstein, Robert, and Gardner (2008b), but we have tried to harmonize our resource representation with subsequent resource ontologies, for example, the BRO (Tennenbaum et al., 2011) and the eagle-i Resource ontology (Torniai et al., 2011). In developing the NIF Resource type module, we in general differentiate between the resource itself and the content/product that it offers. The resource itself is usually identified by a single Web address, for

Table 3.1 List of resources referred to in the text

Resource	Short name	URL
Allen Brain Atlas	ABA	http://mouse.brain-map.org/
Brain Architecture Management System	BAMS	http://brancusi.usc.edu/bkms/
BrainSpan		http://www.brainspan.org/
Brede		http://neuro.imm.dtu.dk/services/brededatabase/
Cell Centered Database	CCDB	http://ccdb.ucsd.edu
Cell Image Library	CIL	http://cellimages.ascb.org/
Collations of Connectivity data on the Macaque brain	CoCoMac	http://cocomac.org/
Gemma		http://www.chibi.ubc.ca/Gemma/home.html
Gene Expression Omnibus:	GEO	http://www.ncbi.nlm.nih.gov/geo/
GeneNetwork		http://www.genenetwork.org/
GeneWeaver		http://geneweaver.org/
Gensat		http://www.gensat.org/
Internet Brain Volume Database	IBVD	http://ibvd.org
Neuroimaging Tools and Resource Clearinghouse	NITRC	http://nitrc.org
Neuromorpho		http://neuromorpho.org
Neuromab		http://neuromab.ucdavis.edu/
Open Access Series of Imaging Studies (OASIS)	OASIS	http://www.oasis-brains.org/
Open fMRI		http://openfmri.org/
Research Portfolio Online Reporting Tools	Nih Reporter	http://reporter.nih.gov
SynapseWeb		http://synapses.clm.utexas.edu/
Surface Management System database	SUMSdb	http://sumsdb.wustl.edu:8081/sums/index.jsp
UCLA Multimodal Connectivity Database		http://umcd.humanconnectomeproject.org

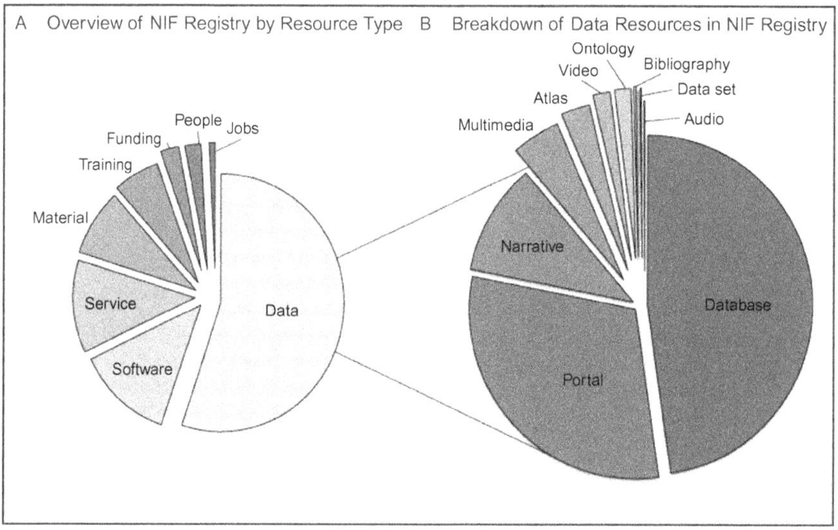

Figure 3.2 NIF Registry Content. (A) Represents NIF Registry content by resource type, while (B) provides an expansion of data resources, to illustrate the diversity of data and information resources available. Some of the smaller categories under data (<25 total) were excluded for clarity, this included license, listserv, thesis, discussion, audio track, bibliography, and slide. (For color version of this figure, the reader is referred to the online version of this chapter.)

example, the Allen Brain Atlas, but may offer several different datasets, products, and services. Each of these is given its own registry entry, but linked to the parent entity.

NIF Data Federation: As shown by the number of resources in the NIF Registry (Fig. 3.1), the number of resources available of potential interest to neuroscience is extremely large. The registry currently lists over 2000 databases. The large number of databases and the difficulty in characterizing their content via a few high-level keywords were the major motivation for the creation of the NIF Data Federation (Gupta et al., 2008). While all resources enter the NIF via the NIF Registry, only a subset of available sources are available via the federation, ~150 as of this writing, although NIF continues to deeply federate resources at the rate of 25–40 per year. These 150 sources collectively comprise >330 million data records. Selection of resources for the federation is driven by a variety of factors including neuroscience relevance, coverage, and willingness of the resource provider to permit access.

Each resource within the federation is characterized roughly by data type and also level of the nervous system (Fig. 3.1). For each federated source,

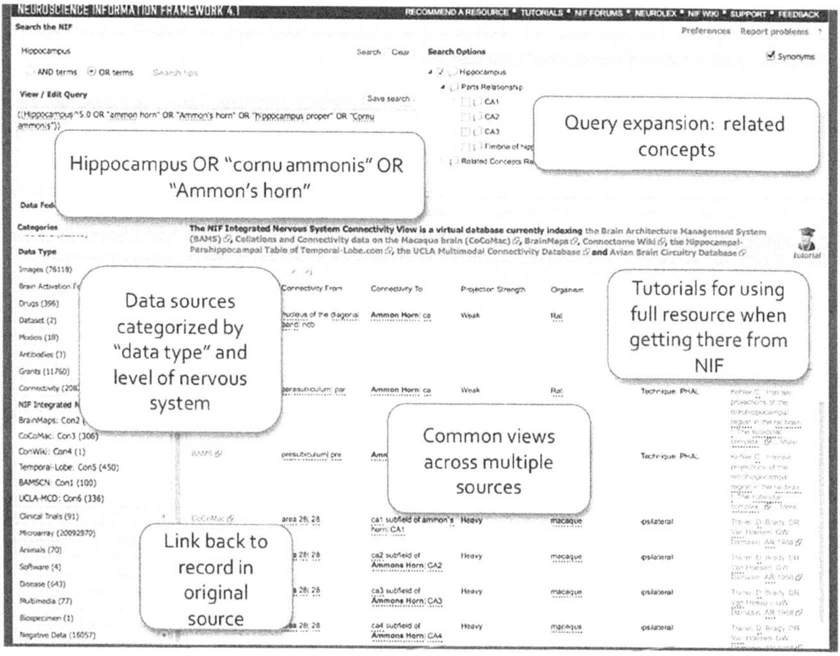

Figure 3.3 Current results display for the NIF Integrated Connectivity data set from the NIF Data Federation for the query "hippocampus." The query automatically searches for synonyms joined by an "OR" operator (*cornu ammonis* or Ammon's horn). The advanced search box on the right provides additional related classes that can be added to the search. The left panel organizes the results retrieved from the federation by data type and level of the nervous system (not shown). Within each category, the individual data sources are displayed, along with the number of records available. For NIF's integrated views, we also display the results available from the individual sources comprising the view. (For color version of this figure, the reader is referred to the online version of this chapter.)

NIF creates a view that provides an overview of the key data contents of the resource (Fig. 3.3). Generally, this view contains a mixture of what would be considered metadata (e.g., subject attributes) and data (i.e., the information object offered by the database). These views are created to allow NIF users to rapidly scroll through the contents of different databases to see what is available and what might be useful to them. For very complex resources, NIF may define multiple views of the contents. Thus, NIF rarely exposes the entire contents of the database or data set through the portal, although a more complete set of is typically available through NIF export services. Most databases are available for export in CSV formats, while a smaller number of databases have licensing restrictions that require NIF to disable data export.

As NIF has developed, we have tried to unify the presentation of results within the Data Federation as much as possible using the semantic framework established via the NIFSTD ontologies. As such, NIF attempts to use a consistent set of labels as column headers for each resource, to make it easier for users to navigate between them. For example, NIF will replace "Species" with "Organism," as that is the root class in the organism module of NIFSTD. In an ongoing process, NIF has been mapping the contents of a given database to the NIF ontologies in order to minimize terminological heterogeneity across and even within sources. As many of these databases were created before there were standard vocabularies or ontologies, most use their own custom terminology. Such concept-mapping efforts do not change the actual data record but provide a unifying semantic layer a top of the original data to improve query and integration. NIF has not tried to solve the resolution among sources at the deep semantic level, that is, does cerebral cortex refer to the same set of substructures in the Allen Brain Atlas (Lein, Hawrylycz, et al., 2007) as it does in the GENSAT atlas (Gong et al., 2003), mostly because this level of concept resolution requires significant effort, particularly where the source database does not provide definitions. Rather, NIF has reconciled resources at the level of a controlled vocabulary, so that synonymous terms like *cornu ammonis* and Ammon's horn are both retrieved via a single query (Fig. 3.3). In this case, NIF's concept-based search, which automatically searches for all synonyms, will retrieve related records.

The bulk of NIF's content to date, however, has not been explicitly mapped to ontology identifiers, unless a custom abbreviation or symbolic notation was used within the database. For example, the SUMSdb of brain activation foci used the custom shorthand Brodmann.3 to denote the cytoarchitectural parcellation of cerebral cortex commonly referred to as Brodmann Area 3. In other cases, databases might use the value 1 to represent male and 2 for female, which would not be understandable outside of the source database context. Thus, NIF replaces these notations with the appropriate term. However, as the content of NIF has grown, the need to disambiguate entities that have similar names across domains has become more acute, so we have initiated the process of mapping all NIF contents to NIFSTD ontologies. As of NIF 4.5 (release in June, 2012), NIF will employ a semi-automated concept-mapping tool of source content based on Google Refine (http://code.google.com/p/google-refine/). Resource providers will be able to use Google Refine on their own resources to map their data to the NIFSTD ontologies, thereby facilitating integration within the NIF and across resources.

The NIF includes two types of views of individual resources within the data federation, which we term vertical and horizontal. Vertical views represent key information from a single source while horizontal views combine similar information from multiple sources. For example, NIF Connectivity combines brain connection statements from six different databases (Fig. 3.3). In these cases, NIF uses its domain expertise to identify commonalities among different datasets that contain essentially the same type of information. In this case, all of the connectivity databases contained pairs of brain regions and a measure of the strength of connection between them. Each, however, represented this information differently, both in terms of data model and in terms of user interface, making it very difficult to compare among them. NIF combined them into a single view where each row links back to the original source database. We are in the process of performing concept-mapping across these views to help unify the terminology to improve analysis of this integrated brain connectivity sources.

Considering the Data Federation as a whole, the largest amount of data, by an order of magnitude, comes from microarray studies, representing a total of nine distinct resources (Fig 3.4A). Microarray resources include general microarray storage repositories, for example, GEO Gene Omnibus, and

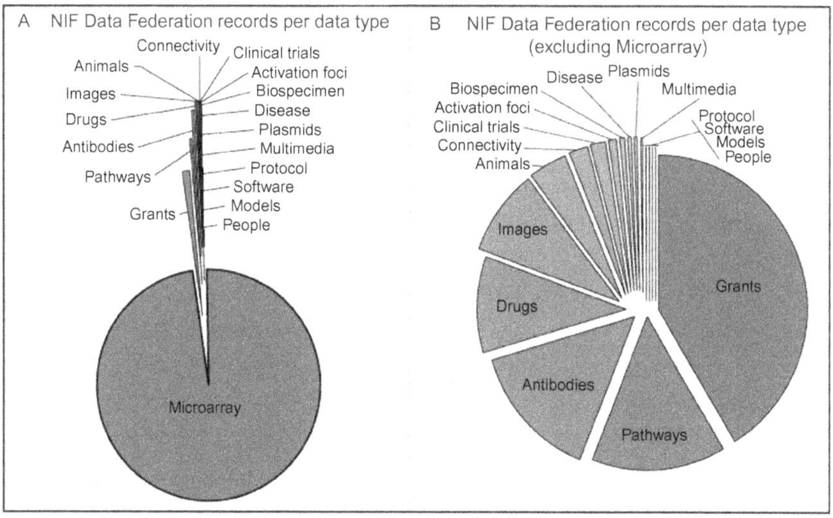

Figure 3.4 NIF Data Federation Content. (A) Provides the percentage of records within the NIF Data Federation, per data type (notice that microarray records dwarf total contents of all other data types combined), (B) represents the percentage of records excluding microarray data. (For color version of this figure, the reader is referred to the online version of this chapter.)

more neuroscience-centric resources, for example, GeneNetwork, Gene Weaver, BrainSpan, Gemma, Drug Related Gene Database (DRG). Many of these data are derived from behavioral experiments, for example, on the effects of drugs of abuse on gene expression. NIF currently presents both primary data repositories, for example, GEO, BrainSpan, and also derived repositories, which offer reanalysis (e.g., Gemma, GeneNetwork, or Gene Weaver) or additional tools for working with these data (e.g., Gene Network).

Excluding microarray datasets, the Data Federation has a diverse array of deeply integrated data types (Fig. 3.4B). Many of the data types are fairly specific for neuroscience, for example, brain connectivity or brain activation foci, while others are more generally relevant, for example, animal models, biochemical pathways. Currently, behavioral data are categorized under the loose term "Nervous system function," comprising 13 different databases of physiological data ranging from cellular models to functional imaging data, Although in the past, few databases were available for behavioral data in the neurosciences, with current emphasis on functional brain imaging in humans, many more such resources are becoming available. For some areas, NIF can certainly claim to have comprehensive coverage, that is, it provides access to a significant portion of data available. These areas include animal models (major model organism databases for worm, rat, mouse) and antibodies (over 900,000 antibodies aggregated from commercial and non-commercial vendors). Other data sets are large, for example, connectivity (132, 700 connectivity statements from 6 databases), but it is difficult to estimate how comprehensive it is relative to all potential sources of information. In the case of connectivity, only one source (the UCLA Multimodal Connectivity Database) contains primary data; the other sources (BAMS, CoCoMac, etc.) are databases that largely contain connectivity statements derived from published studies. As both BAMS and CoCoMac have limited focus (rats in the first case; primate cortex in the latter) and rely on individuals for population, it is likely that the majority of connectivity data available from published studies are not queried directly by the NIF and that the dataset available is highly biased. Indeed, when we plot the number of connectivity statements per brain region, we see that coverage of major brain structures is not uniform (Fig. 3.5). In our current data set, the amygdala and its subnuclei and the cerebral cortex are the most heavily represented structures. Of course, these are richly interconnected structures, but the result set is highly biased toward these structures likely due to the focus of the laboratories originating the BAMS and CoCoMac datasets.

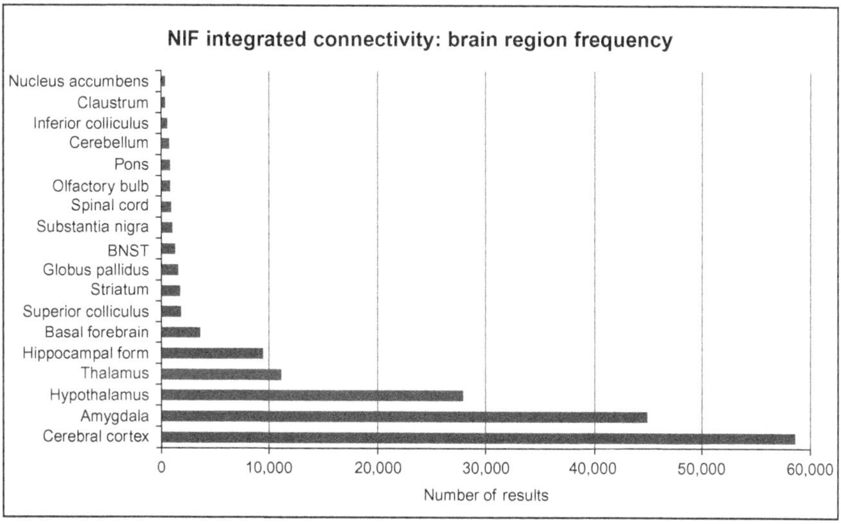

Figure 3.5 NIF Integrated Nervous System Connectivity: Frequency of Brain Region Data. The NIF Integrated Nervous System Connectivity view is a virtual database providing a composite index of five databases: the Brain Architecture Management System (BAMS; http://brancusi.usc.edu/bkms), Collations of Connectivity data on the Macaque brain (CoCoMac; http://cocomac.org), BrainMaps (http://brainmaps.org), ConnectomeWiki (http://www.connectome.ch), Hippocampal-Parahippocampal table of Temporal-Lobe.com (http://www.temporal-lobe.com), the Avian Brain Circuitry Database (http://www.behav.org/abcd/abcd.php), and the UCLA Multimodal Connectivity Database (http://jessebrown.webfactional.com/). This figure reports the number of results returned for each brain region, including their major parts as defined within the NIF ontologies (NIFSTD v2.5/NIF Anatomy v1.3). Within these databases, there are many more connectivity statements regarding the cerebral cortex or amygdala, compared to other regions such as the spinal cord or nucleus accumbens. BNST, Bed nucleus of stria terminalis; Hippocampal form, Hippocampal formation.

NIF also provides access to a large collection of imaging-related data, from microscopy (CCDB, SynapseWeb, Cell Image Library) to brain imaging. As with connectivity, some of these sources represent primary data while other are result sets extracted from the literature. For example, for brain activation foci, NIF searches over two major databases, Brede and SUMSdb, which themselves aggregate brain activation foci from the literature. NIF provides a metadata search capacity over several other functional imaging sources, which then take users to sites where the data may be downloaded, for example, 1000 Functional Connectome datasets, available for download from NITRC, and the open fMRI data repository. In addition, structural brain scans are available via XNAT/OASIS.

3.1. Data, derived data, and metadata

A frequent question to the NIF is whether or not NIF has data as opposed to just providing a deep index over data sources. The brief discussion of the current NIF Data Federation above highlights that while the answer is clearly "yes," data resources themselves are highly diverse and some comments on the nature of NIF's data are warranted. Here, we consider two aspects of data resources apparent from an analysis of the current NIF data federation: (1) types of data; (2) data "liquidity."

3.1.1 Types of data

Even the brief survey presented above suggests that all databases are not the same in terms of the information that they provide, irrespective of any consideration of technology platform. As NIF is charged with surveying the landscape of neuroscience-relevant resources, we take a broad view of what constitutes data. If we consider the contents of the current NIF data federation, we see that there are databases for what we might call "primary" data, that is, the measurements that were taken in the course of a study. Note that the data collection itself need not have come from the same study. For example, the Cell Centered Database, a database of 3D microscopic imaging data, or GEO, present data that was collected from multiple groups, but these resources provide the data products of the study, rather than quantities or qualities derived from them. These types of resources are not restricted to quantitative data alone, as a database like the NIH Grants Reporter, which provides a list of grants awarded by NIH, would also be considered within this category.

Other resources present derived data, defined here as data that was obtained through analysis of the primary data products. Again, there is considerable heterogeneity in these types of databases. In the first case, we have measurements of primary data features based on some additional processing. Examples might include measurements that were taken of brain structures, for example, the Internet Brain Volume Database (IBVD), the Neuromorpho neuronal reconstruction database, or the Allen Brain Atlas, which contains calculations of gene expression per brain region or voxel of brain. Note again that we find sources where the data are aggregated from multiple studies whereas others are single source: the IBVD and Neuromorpho aggregate these quantities from different studies, whereas the Allen Brain Atlas derived these data from the data they generated. However, in both cases, we are presented with a quantity that represents a secondary measurement performed on imaging data. Not all derived data need be quantitative, as one might reasonably claim that the brain region connectivity statements contained in BAMS or CoCoMac are also measurements based on the primary data, even though the measurement in this case is a qualitative statement about perceived presence or absence of a connection. The goal of these derivations is to turn features of the data product into a structured or computable form.

We also see databases that contain another level of derived data. Generally, these fall into the category of claims or assertions about the meaning or significance of data that reflect the results of an experimental paradigm. For example, the claim that gene expression was increased as a function of age (Gemma) or that the hippocampus is activated in verbal fluency tasks (e.g., Brede and SUMSdb). In these cases, a change in value is noted as the result of an experimental analysis. The lines between these two types of claims blur in many instances, as any evaluation of a quantity like labeling intensity implies a comparison to something, even an internal control. Nevertheless, the second type of claim generally has a significance attached to it as the result of a statistical analysis where a difference due to an independent variable is noted, rather than simply an observation or calculation about a data attribute. Again, although these types of claims can be derived from single source studies, generally, the databases that contain them are aggregators, for example, the DRG, SUMSdb, Brede. We also see that the same source may provide both primary and derived data.

In summary, we see from NIF that resources can be grouped roughly into single source versus aggregation databases and primary versus derived data. We also see many instances of what we would call registries, which contain high-level metadata and pointers to information stored elsewhere. Aggregation can be performed at the data set level, for example, GEO or at the individual data point level, for example, CCDB. All of these sources contain metadata that provide key attributes of the subjects, experimental conditions, or data types that are required to understand the context of the data. In general, users can download either the entire data set or a view on the data via the NIF interface or access them through Web services; thus, we can say that NIF hosts data. However, in other cases, NIF only queries the metadata and requires the user to access the original source in order to obtain a copy of the data. Decisions are made based on a consideration of time and effort available both on NIF's and the resource provider's.

3.1.2 Data liquidity

A second feature of data resources that we see clearly within the NIF Data Federation is that the data themselves flow from one resource to the next. This liquidity may simply represent a "pass through" model where the data are hosted by multiple resources, usually for convenience and to achieve improved performance of a particular system. However, in most cases, data are ingested into another resource so that value can be added, either in the form of integrated data or the availability of new analysis tools. These data may be contained within a published article or hosted by a repository like GEO. Of NIF's current data sources, 29 contain explicit per record references to articles within PubMed. NIF provides automatic linkages of these data records to the PubMed record via the NCBI Link Out function (Marenco, Ascoli, Martone, Shepherd, & Miller, 2008). As of this writing, over 900,000 of linkages contained within NIF data sources have been included within PubMed.

For resources that import similar data, keeping track of these external references allows computer systems to calculate degree of overlap between resources more easily. For example, both Brede and SUMSdb aggregate data from published functional imaging studies. Comparing the PubMed ID's from the two databases indicates that 269 papers are common across the two databases. Table 3.2 shows some of the data records available in each database for the study by Phelps et al. (1997) as viewed through the NIF. One can see that the two data sets are not identical in the information they provide, nor in the coordinates given. In this case, SUMSdb adds an additional normalization step that helps to align coordinates across studies, although the original coordinates are available through SUMSdb. SUMSdb provides an additional mapping of the coordinates to Brodmann's areas. By aligning the two data sets, additional value is added not present in either source alone.

A second example is the use of microarray data sets. NIF lists at least four resources that ingest data from GEO and add additional data or analyses to these: Gene Network, Gene Weaver, Gemma, and DRG. DRG was created by NIF as part of a project to survey data contained within tables, figures and supplementary material. DRG includes gene expression data contained within supplementary material, and tables and figures from the literature. It translates gene expression information both from microarray and immunocytochemical experiments. Unlike Gemma, which reanalyzes the data using a custom algorithm (French, Lane, Law, Xu, & Pavlidis, 2009), DRG translates assertions made by the authors regarding differential expression. Many of these assertions are not contained within text but in tables and figures, and so typically are not captured by text mining tools. A human curator performed the translation. To add to the fluidity, GeneWeaver imports result sets from DRG. Some of this content overlaps with content currently in GeneWeaver, for example, datasets GS14912 and GS87486.

Table 3.2 An aggregation of the brain activation foci (x, y, z) results provided by NIF for the SUMSdb (light blue) and Brede (white) databases, extracted from the study of Phelps, Hyder, Blamire, and Shulman (1997) employing a "Generate word beginning with given letter versus simple repetition of heard word" task

SUMS			Brede			SUMS	Brede	SUMS	Brede
X	**Y**	**Z**	**X**	**Y**	**Z**	**Geography**	**Anatomy**	**Area**	**behavioral_domain**
−3	20	42	−4	20	40	Cingulate gyrus/sulcus; LOBE. FRONTAL; SUL.CiS	Cingulate gyrus/sulcus	Brodmann.32	Cognition, Language—Verbal fluency
5	18	27	4	17	27	Cingulate sulcus; LOBE.LIMBIC; SUL	Cingulate sulcus	Brodmann.24	Cognition, Language—Verbal fluency
−47	25	18	−46	24	18	Inferior frontal gyrus; LOBE. FRONTAL; SUL.IFS	Inferior frontal gyrus	Brodmann.45	Cognition, Language—Verbal fluency
−23	28	47	−23	28	44	Middle frontal gyrus/superior frontal sulcus; LOBE.FRONTAL; SUL.SFS	Middle frontal gyrus/superior frontal sulcus	Brodmann.8	Cognition, Language—Verbal fluency

Additional results are available in NIF and in the original sources.

To compare the representations of the same data set in two resources, we searched for GEO datasets that were present in at least two other resources. We found two GEO datasets that were both represented in Gemma and DRG. Both Gemma and DRG contain results of gene expression based on a consideration of the experimental paradigms used. Thus, each presents a statement about differential expression across combinations of groups and conditions. Of these Gemma contained reanalysis results for GEO: GSE7762, from a study by Korostynski, Piechota, Kaminska, Solecki, and Przewlocki (2007) on the effects of morphine on gene expression in the striatum. For this dataset, Gemma provided a list of 8001 comparisons, while DRG contained ~13,000 comparisons. The difference in overall number of records represented in part the difference in focus of the two resources. DRG noted genes that were asserted by the authors to be significantly expressed or unchanged, while Gemma only included a list of genes that were differentially expressed. Part of the difference also arose from the different factoring of the experimental variables by the two resources. This particular study had three treatment groups (chronic cocaine, acute cocaine, saline) and four strains that were studied. The DRG did not completely represent all of the basal expression differences among the strains. Thus, the two data sets only contained a subset of results that could be compared. Both DRG and Gemma use the NIF annotation standard for expression results (http://NeuroLex.org/wiki/Category:NIF_annotation_standard), indicating whether a gene showed increased or decreased expression, making it easy to compare the change in expression across the two resources. However, our initial comparison found the direction of change to be opposite in the two databases. Further analysis indicated that the results presented by Gemma note the difference in expression relative to the experimental group, for example, igrm2 in saline treated animals shows increased expression relative to animals treated with chronic morphine, while the results from DRG present the difference relative to the control group, for example, igrm2 in chronic morphine showed decreased expression relative to control. Comparison was also complicated by the conventions used in each resource for genes. The DRG organized genes by gene name and probe ID, according to what was presented in the paper, whereas Gemma organized the results by gene symbol and gene ID. Thus, although we could retrieve the two sets from the NIF, we had to perform considerable data alignment and translation before we could derive a comparison set.

From the original sets of comparisons, we selected a set of 1370 results in DRG that were stated to be differentially expressed as a function of chronic or acute cocaine. Of these, 617 were confirmed by the analysis done in Gemma. Thus, only half of the original assertions were confirmed by the reanalysis. In this case, the ability to align the two data sets provided an alternate view of the data. We believe that the ability to perform this type of cross-study analysis is one of the unique values of the NIF, providing researchers not only the ability to reuse data but to track the results of these reuses across resources. As can be seen by this example, the use of standard identifiers for data sets, for entities referenced within them (e.g., Gene ID's), standard ways of reporting results (experimental vs. control) and annotation standards, would make the ability to perform such comparisons trivial.

3.2. Resource utilization via the NIF

NIF maintains several types of analytics to track both NIF utilization and the resources available through NIF. These latter tools provide statistics on the most accessed resources and information on updates and literature citations, via the NIF curation pipeline. The top sources accessed for March of 2012 are shown in Table 3.3. The most accessed source by far is the funding opportunity database that currently searches grants.gov. This result suggests that the majority of users of the NIF portal are research scientists. An interesting comparison can be made between two different components of the NIF, the main portal (http://neuinfo.org) and the NeuroLex Wiki (http://neurolex.org). The latter is a semantic wiki platform initially created to help build and maintain the NIFSTD ontologies, but later adapted to house the NIF Resource Registry so that the entries could be easily linked with neuroscience concepts (Larson et al., in preparation). Each concept in the NIFSTD has its own page and unique URL (e.g., http://NeuroLex.org/wiki/Category:cerebellum). Wikis, unlike relational databases, are readily indexed via search engines. Larson et al (in preparation) performed a detailed analysis of Web traffic to the two sites and noted the striking difference between them. Figure 3.6 compares the sources of Web traffic for March 2012, as generated by Google Analytics. Note that the majority of visits to the NIF portal arrive via referrals from other sites, whereas the majority of visits to NeuroLex arrive via Web searches. The searches that lead to NeuroLex are specific neuroscience-related terms, for example, cholinergic neuron, whereas those that lead to NIF are generally informatics-specific, for example, neuroscience database. The amount of traffic to NeuroLex is three times that of the NIF Portal, suggesting that people use the Web for conducting neuroscience-relevant searches, but only a subset are specifically

Table 3.3 Most accessed sources in NIF Data Federation for March 2012

Source	Total searches
Grants.gov/Opportunity	3947
SumsDB/Activation Foci	1073
CCDB/All Information	1047
GENSAT/GENSAT	828
AntibodyRegistry/ABs	635
BrainInfo/Brain Region	545
ResearchCrossroads/Grants	458
ClinicalTrials/ClinTr	409
NIF Integrated Connectivity	381
Drug Related Gene Database/DRG	375
OneMind/BioBanks	350
RePORTER/CurrentNIHGrants	310
NIF Integrated Animals/Available	217
DrugBank/Drugs	197
OMIM/Genes	176
AllenInstitute/MouseBrainAtlas	162
BrainMaps/Atlas	162
BAMS/BrainRegions	133
NeuroMorpho/NeuronInfo	120
NIF Integrated Software/Info	119
NeuronDB/Receptors	113
Gemma/Microarray	111
ModelDB/Models	106
NIF Integrated Podcast/Podcasts	104
AddGene/Plasmids	94
	15,016

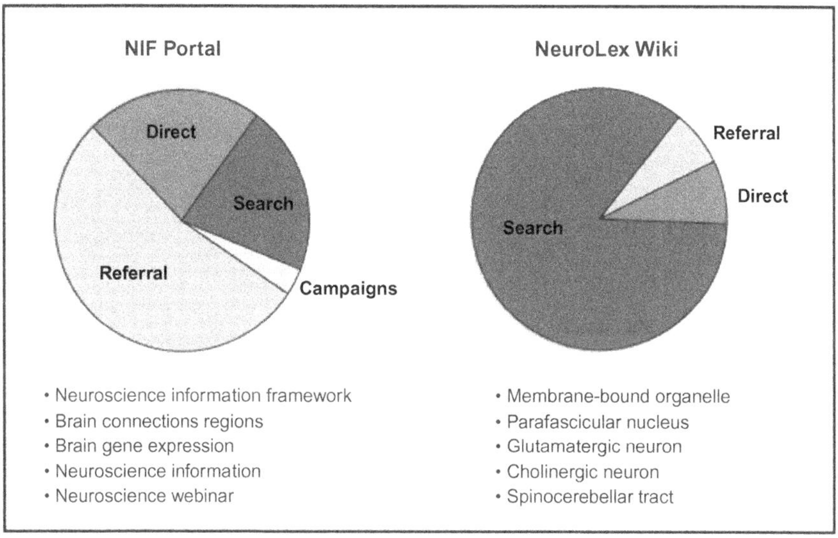

Figure 3.6 Comparison of Web traffic to NIF and NeuroLex. The pie charts show the sources of traffic to the NIF and NeuroLex Web sites (generated by Google Analytics, March 2012). Below each chart are some of the top keywords entered by users that led them to the respective site. (For color version of this figure, the reader is referred to the online version of this chapter.)

looking for data sources. Again, this pattern suggests that those using the NIF portal are primarily research scientists who are looking for data or tools. However, each NeuroLex page contains an embedded NIF Navigator (Fig. 3.1), an applet that searches the NIF for the concept represented by the page. Thus, individuals who search Google for specific neuroscience concepts can query the NIF data federation for additional information. NeuroLex is currently the second largest source of referral traffic to the NIF, suggesting that a subset of users go on to search data sources.

3.3. The NIF resource landscape

The current NIF data set, aggregated largely from a set of metadata and derived data provides a view into the current resource landscape of neuroscience. The semantic framework for understanding this landscape is provided by the NIFSTD, which currently contains >50,000 terms (includes classes and synonyms) from the major domains of neuroscience. The current domains include gross anatomy, cells, subcellular structures, molecules, resources, techniques, dysfunction, and nervous system function. NIF relies on the community to contribute content, either in the form of general community ontologies available via the OBO Foundry (Smith, Ashburner, Rosse, et al., 2007) or more custom ontologies like the Cognitive Paradigm ontology (Turner & Laird, 2012) created for specific neuroscience domains. Analysis of search behavior through NIF shows that of the 7000 or so unique searches through NIF in a month, roughly 4000 are autocompleted via NIF, suggesting that NIF has reasonable coverage of the types of neuroscience concepts used for search. Not all areas, therefore, are equally represented within the NIFSTD, with coverage most extensive for anatomical structures and least for functional entities, like cellular physiology and behavior. As the community develops ontologies in these areas, for example, the Cognitive Paradigm Ontology, NIF imports them. With the continued development of functional imaging to investigate human behavior, more of these types of ontologies are being developed.

To analyze the collective NIF data set, we have recently begun to utilize the Kepler workflow engine (Altintas et al., 2010) to perform custom analytics of NIF data (Astakhov et al., 2012). The heat map, shown in Fig. 3.6 (also see http://www.neuinfo.org/NIF_Federated_Data_Heatmap.html), shows the representation of major brain structures, calculated by searching across the top 3 levels of the NIF Anatomy module, in the result set for each resource in the data federation, organized alphabetically. This visualization shows the neuroscience-centric content of the NIF data federation, as the term "brain" occurs in the majority of data sources. Not surprisingly, the most extensive coverage is found for major brain areas such as cerebral cortex, thalamus, olfactory bulb, hypothalamus, and striatum (Fig. 3.7). The vertical axis clearly shows the resources with the broadest coverage of neuroanatomical structures, suggesting that these are highly neuroscience specific.

3.4. Discussion

The NIF project was initiated to address the breadth and depth of electronic resources available to neuroscientists. As the NIF has grown, it has not only accumulated a significant catalog of what is available but also acquired a global view of data and data resources that examines resources not in terms of what they are but how they can be fit into a neuroscience-centered information framework. NIF specifically addresses the "long tail of small data," aggregating together the sum total of resources available, whether produced by an individual laboratory, NCBI, or the Allen Brain Institute. If one considers the latent complexity of biological systems and the difficulty in interrogating any but a small piece of them at any one time, we can reasonably

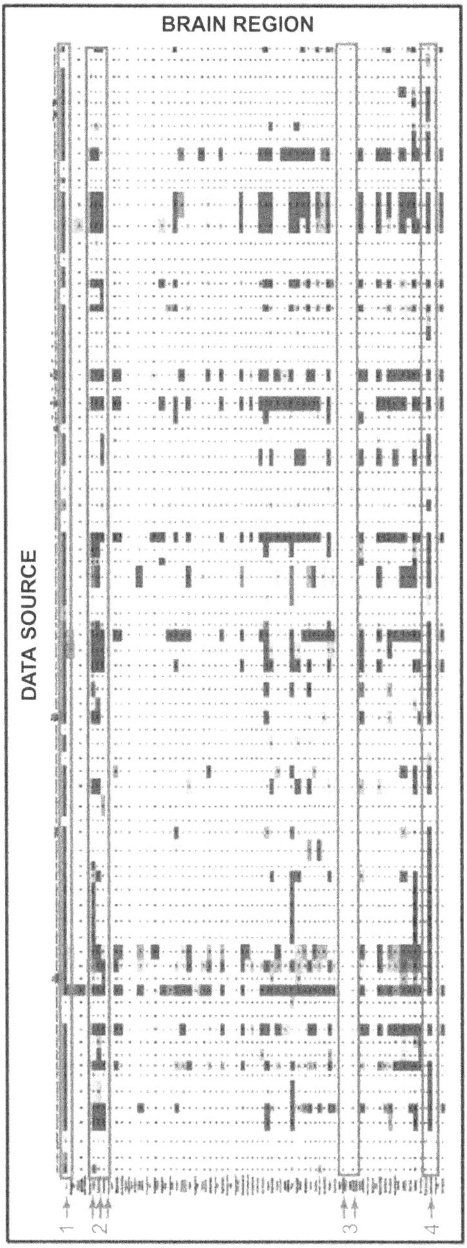

Figure 3.7 Analysis of brain region representation in NIF Data Federation. In this table, each data source in the NIF Data Federation is represented in a column, and each row contains a brain region of interest. This heat map landscape analysis permits a rapid assessment of the overall representation a brain region receives throughout the content of the NIF Data Federation. The darker colors denote more hits, or matches for that brain region within the respective data source. For example, regions marked 1 (brain), 2 (striatum, hypothalamus, olfactory blub), and 4 (cerebral cortex) are well represented in almost all data sources. However, regions marked 3 (pontine tegmentum, ventral amygdalofugal projection) have almost no associated content. (For color version of this figure, the reader is referred to the online version of this chapter.)

state that as far as neuroscience is concerned, there are only small data. That is, no single technique or resource to date holds the entire key to unlocking the secrets of the brain. With the buzz surrounding big data analytics, NIF hopes to help inculcate within the biomedical research community a similar global perspective on data that will lead to building of resources and reporting of scientific data in a manner that makes it easier to aggregate them within the framework. From NIF's perspective, sharing data requires that we can (1) find them, (2) access them, and (3) understand enough context to use them.

The NIF Resource Registry and Data Federation collectively represent one of the largest collections of biomedical resources available on the Web. As such, they provide a means to assess the current landscape of biomedical resources. Not surprisingly, we see quite a few projects that are similar in scope and stated goals. Databases are developed that contain largely the same type of content, sometimes even with overlapping content. As our continual surprise at the discovery of significant new resources over the course of the NIF project has shown, some databases may be duplicated simply because of ignorance of the other efforts. Databases may also be duplicated because they have a slightly different focus, or believe they have an improved representation, tool set, or quality compared to an existing resource. Multiple efforts may be launched around the same time around new technologies. An entire issue of *NeuroImage* was devoted to the topic of brain activation foci representation within databases, and a brief perusal of the commentaries suggest that the community is far from in agreement as to the best way to make brain activation foci searchable (e.g., Derrfuss & Mar, 2009). Given the way that biomedical science is funded, the intense competition among scientists and the lack of incentives for contributing to community resources, NIF believes that some duplication is inevitable. But, as we also show here, this duplication can be used to advantage in that it provides some means to aggregate information, assess the effectiveness of different representations, and even the reproducibility of data results. However, this advantage cannot be realized if we lack effective means to aggregate and compare these data sets across resources.

NIF has continually added content to both to the registry and the data federation since the first production release in 2008. In retrospect, we can clearly see different stages in data ingestion over that time period. The initial period focused on cataloging and surveying available resources (Gardner et al., 2008a). The next phase focused on developing the semantic framework and technologies for providing deep search across independent databases, ensuring that we could ingest sources based on different technological platforms and across diverse domains within neuroscience and effectively search them (Bug et al., 2008; Gupta et al., 2008; Imam et al., 2012). As the NIF data federation became populated, the next phase focused on providing more unified views of these resources to make them easier to understand through NIF and to compare with one another. Initially, this work focused on the production of the horizontal views across similar sources and providing a more uniform look and feel to data within the NIF portal. The completion of this phase will be realized with the release of NIF 4.5 in summer of 2012, which will largely complete the mapping of terminologies to the NIFSTD using Google Refine.

The current phase focuses more on the linkages across data, and providing a unified view of the NIF resource landscape so that these linkages are apparent. The evolution of NIF mirrors the nature of data resources themselves and highlights the difference between databases and publications. As data flows from one application to another, it becomes transformed as new annotations are added, new information is derived from them, and additional data are aggregated to them. But unlike the publication, where there is an enduring artifact that can be referenced, the issue of identifying data has proven more challenging. In NIF's current phase, we are focusing on establishing effective means to show the interconnectedness of our data sources, by exposing external references like GEO ID's or PubMed ID's in a more uniform manner. Toward this end, the NIF is now including the identifiers of any external reference in all views of data available through the NIF. We strongly encourage resource providers to include these ID's in their resources, rather than textual citation information. Ironically, however, just as with terminology, the heterogeneity of external references can present problems for effective search and integration. Even a standard ID such as an ontology ID or a data set reference ID can be presented in multiple ways, leading to false negatives. For example, some resources prepend the source to the ID, for example, GEO:GSE7762, while others just present the GSE number in a column entitled "GEO ID". Several groups are working to define standards for data reference, e.g., BioDBcore (Gaudet et al., 2011) and http://Identifiers.org that will provide standard references for data. By using a standard reference, searching the NIF for a PubMed or GEO ID will bring back all references to that data within the NIF data federation.

The value of these resources and aggregations produced from the long tail of small data is difficult to predict, as we still learning to extract information from messy, heterogeneous data sets. We can see, however, that scientists are producing different types of information entities, beyond simple publications, that attempt to make sense out of the mounds of data available. The NIF performs a service by allowing these different types of entities to be collectively searchable, much in the way we can search across all Web documents or biomedical abstracts. What is also clear, even from this limited survey of the resource landscape, is that viewing the collective output of the scientific community as part of a virtual global repository, rather than an isolated piece of information, helps us ask additional types of questions beyond their original purpose. As highlighted in a recent editorial (Begley & Ellis, 2012) bemoaning the lack of reproducibility of basic scientific findings, "The scientific community assumes that the claims in a preclinical study can be taken at face value-that although there might be some errors in detail, the main message of the paper can be relied on and the data will, for the most part, stand the test of time. Unfortunately, this is not always the case." By developing community platforms for publishing data and not just narrative, as well as platforms like NIF for accessing them and facilitating their use, we believe that the process of science will be improved, and that insights can be gained through query over the entire data landscape.

ACKNOWLEDGMENT

Supported for NIF is provided by a contract from the NIH Neuroscience Blueprint HHSN271200800035C via the National Institute on Drug Abuse.

REFERENCES

Akil, H., Martone, M. E., & Van Essen, D. C. (2011). Challenges and opportunities in mining neuroscience data. *Science*, *331*(6018), 708–712. http://dx.doi.org/10.1126/science.1199305.

Altintas, I., Lin, A. W., Chen, J., Churas, C., Gujral, M., Sun, S., et al. (2010). CAMERA 2.0: A Data-Centric Metagenomics Community Infrastructure Driven by Scientific Workflows. In: *SWF 2010 in conjunction with 6th World Congress on Services (SERVICES 2010)*, pp. 352–359.

Astakhov, V., Bandrowski, A., Gupta, A., Kulungowski, A. W., Grethe, J. S., Bouwera, J., et al. Prototype of Kepler processing workflows for Microscopy and Neuroinformatics, International Conference on Computational Science, ICCS 2012, Procedia Computer Science (http://www.sciencedirect.com/science/article/pii/S1877050912002967).

Bandrowski, A. E., Cachat, J., Li, Y., Muller, H. M., Sternberg, P. W., Ciccarese, P., et al. (2012). A hybrid human and machine resource curation pipeline for the Neuroscience Information Framework. *Database*, http://dx.doi.org/10.1093/database/bas005.

Barnes, S. J., & Shaw, C. D. (2009). BrainFrame: A knowledge visualization system for the neurosciences. *Proc. SPIE 7243,* Visualization and data analysis 2009, 72430F (January 18, 2009); http://proceedings.spiedigitallibrary.org/proceeding.aspx?articleid=812184. http://dx.doi.org/10.1117/12.812290.

Begley, C. G., & Ellis, L. M. (2012). Drug development: Raise standards for preclinical cancer research. *Nature*, *483*(7391), 531–533. http://dx.doi.org/10.1038/483531a.

Bergman, M. K. (2001). White paper: The deep web: Surfacing hidden value. *Journal of Electronic Publishing*, 7(1), 1–17. http://dx.doi.org/10.3998/3336451.0007.104.

Bug, W. J., Ascoli, G. A., Grethe, J. S., Gupta, A., Fennema-Notestine, C., Laird, A. R., et al. (2008). The NIFSTD and BIRNLex vocabularies: Building comprehensive ontologies for neuroscience. *Neuroinformatics*, *6*(3), 175–194.

Derrfuss, J., & Mar, R. A. (2009). Lost in localization: The need for a universal coordinate database. *NeuroImage*, *48*(1), 1–7.

French, L., Lane, S., Law, T., Xu, L., & Pavlidis, P. (2009). Application and evaluation of automated semantic annotation of gene expression experiments. *Bioinformatics*, *25*(12), 1543–1549.

Gardner, D., Akil, H., Ascoli, G. A., Bowden, D. M., Bug, W., Donohue, D. E., et al. (2008). The Neuroscience Information Framework: A data and knowledge environment for neuroscience. *Neuroinformatics*, *6*(3), 149–160.

Gardner, D., Goldberg, D. H., Grafstein, B., Robert, A., & Gardner, E. P. (2008). Terminology for neuroscience data discovery: Multi-tree syntax and investigator-derived semantics. *Neuroinformatics*, *6*(3), 161–174.

Gaudet, P., Bairoch, A., Field, D., Sansone, S. A., Taylor, C., Attwood, T. K., et al. (2011). Towards BioDBcore: A community-defined information specification for biological databases. *Database (Oxford)*, baq027.

Gong, S., Zheng, C., Doughty, M. L., Losos, K., Didkovsky, N., Schambra, U. B., et al. (2003). A gene expression atlas of the central nervous system based on bacterial artificial chromosomes. *Nature*, *425*(6961), 917–925.

Gupta, A., Bug, W., Marenco, L., Qian, X., Condit, C., Rangarajan, A., et al. (2008). Federated access to heterogeneous information resources in the Neuroscience Information Framework (NIF). *Neuroinformatics*, *6*(3), 205–217.

Hey, A. J., Stewart, T., & Kristin, M. (2004). *The Fourth Paradigm: Data-intensive Scientific Discovery*. Redmond, WA: Microsoft Research.

Imam, F. T., Larson, S., Grethe, J. S., Gupta, A., Bandrowski, A., & Martone, M. E. (2012). Development and use of ontologies inside the Neuroscience Information Framework: A practical approach. *Frontiers in Bioinformatics and Computational Biology*, (accepted pending revision).

Korostynski, M., Piechota, M., Kaminska, D., Solecki, W., & Przewlocki, R. (2007). Morphine effects on striatal transcriptome in mice. *Genome Biology*, *8*(6), R128.

Kötter, R. (2001). Neuroscience databases: Tools for exploring brain structure-function relationships. *Philosophical Transactions of the Royal Society of London. Series B, Biological Sciences*, *356*(1412), 1111–1120. http://dx.doi.org/10.1098/rstb.2001.0902.

Lein, E. S., Hawrylycz, M. J., et al. (2007). Genome-wide atlas of gene expression in the adult mouse brain. *Nature*, *445*(7124), 168–176.

Marenco, L., Ascoli, G. A., Martone, M. E., Shepherd, G. M., & Miller, P. L. (2008). The NIF LinkOut broker: A web resource to facilitate federated data integration using NCBI identifiers. *Neuroinformatics*, *6*(3), 219–227.

Marenco, L., Wang, R., Shepherd, G. M., & Miller, P. L. (2010). The NIF DISCO Framework: Facilitating Automated Integration of Neuroscience Content on the Web. *Neuroinformatics*, *8*(2), 101–112.

Martone, M. E., Gupta, A., & Ellisman, M. H. (2004). E-neuroscience: Challenges and triumphs in integrating distributed data from molecules to brains. *Nature Neuroscience*, 7(5), 467–472.

Phelps, E. A., Hyder, F., Blamire, A. M., & Shulman, R. G. (1997). FMRI of the prefrontal cortex during overt verbal fluency. *NeuroReport*, *8*(2), 561–565.

Smith, B., Ashburner, M., Rosse, C., et al. (2007). The OBO Foundry: Coordinated evolution of ontologies to support biomedical data integration. *Nature Biotechnolgoy*, *25*, 1251–1255.

Tenenbaum, J. D., Whetzel, P. L., Anderson, K., Borromeo, C. D., Dinov, I. D., Gabriel, D., et al. (2011). The Biomedical Resource Ontology (BRO) to enable resource discovery in clinical and translational research. *Journal of Biomedical Informatics*, *44*(1), 137–145. Epub 2010 Oct 16.

Torniai, C., Brush, M., Vasilevsky, N., Segerdell, E. J., Wilson, M., Johnson, T., et al. (2011). *Developing an Application Ontology for Biomedical Resource Annotation and Retrieval: Challenges and Lessons Learned. Proceedings: International Conference on Biomedical Ontology, Buffalo, NY.*

Turner, J. A., & Laird, A. R. (2012). The cognitive paradigm ontology: Design and application. *Neuroinformatics*, *10*(1), 57–66.

CHAPTER FOUR

The Neurobehavior Ontology: An Ontology for Annotation and Integration of Behavior and Behavioral Phenotypes

Georgios V. Gkoutos[*,†,1], **Paul N. Schofield**[‡], **Robert Hoehndorf**[*]

[*]Department of Genetics, University of Cambridge, Cambridge, UK
[†]Department of Computer Science, University of Aberystwyth, Old College, Aberystwyth, UK
[‡]Department of Physiology, Development and Neuroscience, Downing Street, Cambridge CB2 3EG, UK
[1]Corresponding author: e-mail address: geg18@aber.ac.uk

Contents

Abstract

In recent years, considerable advances have been made toward our understanding of the genetic architecture of behavior and the physical, mental, and environmental influences that underpin behavioral processes. The provision of a method for recording

International Review of Neurobiology, Volume 103
ISSN 0074-7742
http://dx.doi.org/10.1016/B978-0-12-388408-4.00004-6

behavior-related phenomena is necessary to enable integrative and comparative analyses of data and knowledge about behavior. The neurobehavior ontology facilitates the systematic representation of behavior and behavioral phenotypes, thereby improving the unification and integration behavioral data in neuroscience research.

1. INTRODUCTION

The study of the behavior of organisms forms a major biological discipline encompassed via the investigation of physical, mental, and environmental influences that underpin behavioral-related processes. Geneticists have been studying behavior since 1800s when Francis Galton started investigating heredity and human behavior systematically (Rose & Rose, 2011). We now know that one of the most important factors for behavioral variation within and across organisms lies in genetic diversity (Hamer, 2002; Mackay, 2008). Behavioral geneticists attempt to unravel this behavioral variation by investigating the underlying mechanisms that govern it in an effort to elucidate our understanding of the pathogenesis of neuropsychiatric disorders (Congdon, Poldrack, & Freimer, 2010).

The great successes and advances both in genomics and in our abilities to quantify and analyze genomic information have transformed genetics over the past decade. Behavioral geneticists take advantage of these in order to gain an in-depth understanding of the genetic architecture of behavior. They seek to understand what genes affect behavior, how they interact with other genes, what is the molecular basis of their allelic variation, and how this variation behaves with respect to the environment (Holden, 2001). One of the tools that they employ to achieve these goals is the use of animal models that provide a platform where complex behaviors can be studied and quantified with substantial progress over the past in recent years, especially in respect with research related to the mouse and the fruit fly *Drosophila* (Mackay, 2008; Wehner, Radcliffe, & Bowers, 2001).

Animal models have been proven useful for unveiling the genetic basis of many behavior-related diseases including various neurodegenerative disorders such as Parkinson's, Huntington's, spinocerebellar ataxia, and Alzheimer's disease, as well as for providing the medium for novel drug discovery. Furthermore, animal models for diseases whose indicators are formed by behavioral observations rather than definitive neuropathological markers are being developed. For example, there are various mouse models of loss of *Fragile X mental retardation 1* (*Fmr1*) or *methyl-CpG-binding protein-2* (*Mecp2*) or *ubiquitin protein ligase 3A* (*Ube3A*) function that underlie syndromes associated with autistic-like behavior (Moy & Nadler, 2007).

There are now large international projects, consortia, and individual labs around the world that study and record the effect of genetic variations in various species and at various levels of granularity. Behavioral screens are part of the assays performed and include the study of a variety of behavioral phenotypes such as reproductive behavior, learning and memory, feeding behavior, sleep, and circadian rhythm (Brown, Chambon, de Angelis, & EumorphiaConsortium, 2005; Levin & Cerutti, 2009; Sokolowski, 2001; Spuhler, 2009; Tecott & Nestler, 2004). The resulting data provide us with a wealth of information that can be exploited to investigate and reveal the molecular basis of behavior and behavioral disorders. However, while other domains of biology have made significant progress in systematically structuring and analyzing their data, we do not currently have a standardized way to characterize behavioral processes and phenotypes (Congdon et al., 2010). The provision of a method for recording behavior-related phenomena is necessary to enable integrative and comparative analyses of data and knowledge about behavior (Gkoutos, Green, Mallon, Hancock, & Davidson, 2004b).

In other areas of biomedical science, similar demands have lead to the generation of various resources that allow for the systematic characterization, organization, and recording of knowledge and data (Schofield, Sundberg, Hoehndorf, & Gkoutos, 2011a). In particular, the advent of the gene ontology (GO) (Ashburner et al., 2000) has provided a critical landmark in the use of ontologies to harmonize the description of domains of knowledge and facilitated the development of several other ontologies for various different domains. Ontologies are structured, standardized terminologies in which some aspect of the meaning of terms has been rendered computable. For example, the GO does not only include textual definitions of its terms but also machine readable, computable relations (such as **is-a**, **part-of**, and **regulates**) that enable the automated traversal of the ontology and analysis of the underlying data. Perhaps more importantly, the standardization of the terminology and the inclusion of computable definitions paved the way for interoperability between biomedical databases and have lead to the possibility for large-scale integration of biomedical data (Bada et al., 2004; Chen et al., 2012, Hoehndorf, Dumontier, & Gkoutos, 2012; Hoehndorf, Dumontier, et al., 2011; Hoehndorf, Schofield, & Gkoutos, 2011).

Here, we present our efforts toward creating a framework that allows the systematic representation of behavior processes and related phenotype manifestations offering the tantalizing possibility of unifying behavioral data across species integrating and translating our knowledge so as to provide new grounds for targeting behavior-related diseases.

2. RESULTS

2.1. Neurobehavior ontology

Understanding what constitutes behavior will depend on its formal definition and the systematic representation of the processes involved in behavioral mechanisms. According to Tinbergen (1963), behavior biology is primarily concerned with four major questions: causation (mechanism), development (ontogeny), function (adaptation), and evolution (phylogeny) (Adcock, 2001). These four questions can be collapsed into two categories—the *proximate* ("how") category that includes causation and development and the *ultimate* ("why") category that includes function and evolution (Bolhuis & Giraldeau, 2009). Although behavior, as a scientific domain, is usually well understood by most behavioral biologists, a clear definition and delineation of the field have been the subject of many scientific debates in the field of behavioral biology and behavioral genetics (Bolhuis & Giraldeau, 2009).

Perhaps this issue is highlighted by the variety and diversity of definitions of behavior. The definitions of "behavior" include:

- ". . .the study of causation of animal movement with respect to all levels of integration" (Tinbergen, 1963),
- "Behavior is characterized by entropic and energetic transductions by an organism, in which the long-term averages convert high entropic and low energetic sensory inputs into low entropic and high energetic outputs" (Hailman, 1977),
- "Behavior is all observable or otherwise measurable muscular and secretory responses (or lack thereof) and related phenomena in response to changes in an animal's internal or external environment" (Grier & Burk, 1992), and
- "A response to external and internal stimuli, following integration of sensory, neural, endocrine, and effector components. Behavior has a genetic basis, hence is subject to natural selection, and it commonly can be modified through experience" (Starr & Taggart, 1998).

Within the context of the work described here, we aim at providing a consistent representation of the behavior domain that can be applied for the annotation of animal experiments and human phenotypes, disorders and diseases. Such a unifying representation framework will permit the integration of data about behavior and behavioral phenotypes recorded across multiple species. For the purpose of building this framework, we understand behavior to be the response of an *organism* or a group of *organisms* to external or internal *stimuli*.

The neurobehavior ontology (NBO) consists of two main components, an ontology of behavioral processes and an ontology of behavioral phenotypes. The behavioral process branch of NBO contains a classification of behavior processes complementing and extending GO's process ontology. The behavior phenotype branch of NBO consists of a classification of both normal and abnormal behavioral characteristics of organisms. A large portion of these characteristics is based on behavioral processes.

Currently, NBO includes 763 terms, over 75% of which have textual definitions, and over one-third have computable definitions that can be used by reasoners for automated classification. Each class is in the neurobehavior namespace and is uniquely identified by a URI of the form: http://purl.obolibrary.org/OBO/NBO_nnnnnnn. The main ontology is available in both the OBO Flatfile Format (Horrocks, 2007) and the Web Ontology Language (OWL) (Grau et al., 2008) on our project Web site which can be reached at: http://behavior-ontology.googlecode.com.

NBO contains relationships and other logical axioms that reference other ontologies, such as GO (Ashburner et al., 2000), Uberon (Mungall, Torniai, Gkoutos, Lewis, & Haendel, 2012), and PATO (Gkoutos, Green, Mallon, Hancock, & Davidson, 2004a). To make a connection between these ontologies and NBO, we use a set of relationships (described below). For example, for the connections between NBO and Uberon, we employ the **by-means** relation.

2.2. Behavioral process ontology

The *Behavioral process* (NBO:0000313) branch of NBO consists of a classification of processes in which a whole organism or a group of organisms is involved. NBO's process branch explicitly extends GO's classification of behavioral processes, and the top-level class *Behavioral process* is related to GO's *Behavior* class (GO:0007610) (using a cross-reference statement).

The upper-level distinctions in the behavioral process branch of NBO are organized by the nature of the processes. For example, high-level classes in the *Behavioral process* branch of NBO include:

- *Kinesthetic behavior*: behavioral processes that are related to movement of the body's muscles, tendons, and joints. These processes are further

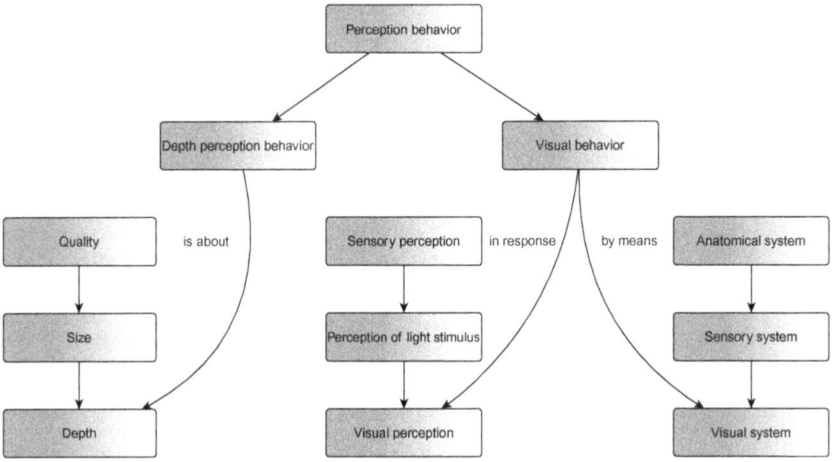

Figure 4.1 Schematic representation of NBO's axes of classification.

distinguished into *Involuntary movement behavioral* and *Voluntary movement behavioral* related processes with subclasses such as *Locomotor activation* and *Body part movement.*

- *Motivation behavior*: behavioral processes that are related to the tendency of an organism to maintain internal equilibrium. Subclasses of this class include *Avoidance behavior, Thirst motivation behavior, Thermoregulation behavior,* etc.
- *Social behavior*: behavioral processes that occur predominantly, or only, in individuals that are part of a group. Subclasses include *Agonistic behavior, Communication behavior, Group behavior,* etc.
- *Cognitive behavior*: behavioral processes that are related to cognition. Examples of process that are categorized here are *Learning behavior, Sensation behavior,* etc.

NBO follows three main axes of classification within its process branch (Fig. 4.1).

First, processes are categorized based on the phenomena to which they are a response. In particular, as we treat behavior as a response of an organism (or a group of organisms) to a stimulus, a natural axis of classification is based on the stimulus to which the organism responds. Formally, we introduce the relation **in-response-to** and use it in axioms that restrict behavioral processes to represent these links computationally. For example, we employ this relation to relate the NBO term *Nociceptive behavior* (NBO:0000331) with the GO term *Detection of electrical stimulus involved in sensory perception of pain* (GO:0050967) in order to formally describe *Chemical nociceptive behavior* (NBO:0000333).

A second axis of classification is based on *intentionality* of behavior. Intentionality is the capability of a mind to represent, stand for, be about or be directed toward something (Searle, 1997). For example, physical symbols (such as "dog") can be observed and interpreted by organisms to *stand for* something else (e.g., the concept *Dog*). Similarly, dreams and hallucinations are *of* or *about* something, emotions (such as fear or love) can be directed *toward* something. Likewise, aggression could be directed *toward* another male organism (*Aggressive behavior toward males* (NBO:0000118)) or a female organism (*Aggressive behavior toward females* (NBO:00000117)) or even on oneself (*Autoaggressive behavior* (NBO:00000742)). For computational access to these relations, we use the **is-about** relation and relate, for example, the *Sensation behavior* (NBO:0000308) with the PATO term *Shape* (PATO:0000052) in order to formally describe *Form perception behavior* (NBO:0000465).

The third axis of classification is based on the means that are used to respond to a stimulus. Some behavioral processes require some means to be performed or some tools to achieve a particular goal, and the *means* axis of classification distinguishes processes based on the means that are used. For example, the NBO term *Behavioral control of lacrimation* (NBO:0000042) represents a behavior related to the regulated release of the aqueous layer of the tear film from the lacrimal glands. To allow computational access to these relations, we use the **by-means-of** relation and relate *behavioral control of external secretion* (NBO:0000041) with the Uberon term *lacrimal gland* (UBERON:0001817). We further employ the **is-about** relation to relate the *behavioral control of external secretion* (NBO:0000041) with the GO term *tear secretion* (GO:0070075). Table 4.1 provides a list of important relations employed by NBO along with their definitions.

Table 4.1 Important NBO relations

Relation	Definition	Example
In-response-to	The relation **in-response-to** holds between a process x and a process y if and only if x occurs in response to y.	A *perception of visual stimulus* process occurs in response to the reception of light in the eye.
By-means-of	A process x occurs **by-means-of** a material structure y if and only if x occurs by means of y.	A perception of visual stimulus process occurs by means of the visual system.
Is-about	A process x **is-about** some entity y if and only if x is about or directed toward y.	A depth perception of process is about depth.

2.3. Behavior phenotype ontology

Phenotypes are observable characteristics of an organism and include characteristics of organism qualities, parts, functions, tendencies, and processes (Hoehndorf, Oellrich, & Rebholz-Schuhmann, 2010). Within NBO, the majority of phenotypes are phenotypic manifestations that are based on the processes in NBO's behavioral process branch. We distinguish between two main types of phenotypes with respect to these processes. Our first main distinction is *single occurrences* of a kind of behavioral process. For all such processes, duration and its deviations (increased/decreased) form a common characteristic. For example, an organism may exhibit *prolonged grooming*. We define such a phenotype as a phenotype of an organism that **participates-in** a *Grooming behavior* (NBO:0000027) that lasts longer than normal, that is, the organism has an *Increased duration* (PATO:0000498) of *Grooming behavior* (NBO:0000027) phenotype.

One other type of observations, we might want to refer to is manifestations that are related to attributes of the process participants in relation to the duration of the processes. For example, it is quite common for behavioral scientists to record the *liquid intake in a single drinking act* (Gooderham, Gagnon, & Gill, 2004). What is intended with such observations is to denote deviations (increased/decreased) of the amount of liquid substance that is consumed. To facilitate annotations, the behavioral phenotype branch of NBO is intended to hold such descriptions. An example would be the NBO term *Increased amount of liquid in a single drinking act* (NBO:00000851) defined as a phenotype of an organism that **participates-in** an *Drinking behavior* (NBO:0000064) that **has-input** some *Liquid* that **has-quality** *Increased mass* (PATO:0001563).

The second major distinction we make is between phenotypes that relate to patterns of *multiple occurrences* of a kind of process. According to GO, regulation processes maintain or modify the occurrence of processes of a particular type. In order to describe behavioral phenotypes of this kind, we describe the phenotypic attributes of regulatory processes. One type of phenotype of regulatory processes is related to their distribution patterns, for example, their frequency. For example, the pattern of frequency of drinking would be an essential characteristic of behavioral phenomena such as *dipsosis* or *hyperdipsia*. For these cases, we describe the phenotype of an organism that **participates-in** a regulation of a *Drinking behavior* (NBO:0000064) with *Increased frequency* (PATO:0000380). We can then use the PATO temporal qualifiers, *Chronic* (PATO:0001863) and *Temporally extended* (PATO:0001333), to distinguish between the two observations.

Further characteristics relate to deviations for behavioral processes' distribution patterns, such as characteristics relating to their rhythm. An example would be *Sleeping behavior* (NBO:0000025), where *Sleep* (GO:0030431) occurs in a rhythmic pattern dictated by *Circadian rhythm* (GO:0007623). Examples of disruptions of such distribution patterns would be circadian rhythm sleep disorders such as *Advanced sleep phase syndrome* or *Jet lag* (Barion, 2011).

Another type of regulatory process phenotypes would be related to their onset. For example, an observation of *Delayed offspring retrieval* would refer to a deviation of the regulation of *Offspring retrieval* (NBO:0000155) in that it is induced later. Such terms could be defined via linking them to the PATO quality *Onset* (PATO:0002325) and its children. The last type of distinction we make refers to the rate with respect to a participant of the process that is being regulated. For example, *polydipsia* refers to an increased amount of liquid intake over a prolonged period of time. This, in contrast to the phenotypes of *hyperdipsia* and *dipsosis* described earlier, does not refer to an *Increased amount of liquid in a single drinking act* rather to an *Increased amount of liquid in drinking regulation* (NBO:0000086) that is prolonged.

In particular, the observation of increased rates of process occurrence is often indicative of an increased *tendency* toward the occurrence of certain processes. For example, from an increased rate of occurrence of aggressive behavior, an inference about an increased tendency toward aggressive behavior can sometimes be made. Although the distinction between both can be relevant in some applications, we do not currently make it explicit in NBO.

In order to capture the differences between these phenotypes, we use the PATO framework (Gkoutos et al., 2004a). According to the PATO framework, phenotypes can be decomposed into the *entities* that have been affected in a particular phenotypic manifestation, which could either be physical objects such as anatomical parts, and the manner that these entities have been affected which are formally termed as *qualities*. PATO supports the use of both pre- and postcomposed phenotype statements. In a precomposed phenotype term, a single term is formally characterized by an entity and a quality, and an annotation is made using the single phenotype term. In postcomposed phenotype terms, data is annotated with multiple terms (i.e., a quality and one or more entities) (Mungall et al., 2010). NBO's *Behavioral process* branch can be used directly with the PATO ontology of qualities in order to describe behavioral phenotypes in a postcomposition manner. NBO's *Behavioral phenotype* branch provides a collection of *precomposed* terms that can be used directly for annotation whilst providing links to affected processes and their qualities thereby ensuring compatibility with postcomposed phenotypic statements.

2.4. Use case: Increased drinking behavior

A phenotype statement such as *Increased drinking behavior* is used for the description of phenotypes in mouse and other species including several human diseases. However, based on the context, *Increased drinking behavior* may have several different meanings, and serve as an example of distinctions that we intend NBO to be able to express. *Increased drinking behavior* may refer to a state in which, for example:

(a) the amount of substance that is consumed is increased over a fixed period of time (e.g., 24 h) (Gooderham et al., 2004),

(b) the amount of substance per drinking act is increased (Gooderham et al., 2004),

(c) the amount of time that is being spent drinking within a fixed period of time is increased (Wood et al., 2008),

(d) the amount of time spent per drinking act is increased (Wood et al., 2008),

(e) the number of drinking acts per fixed period of time is increased (Wood et al., 2008),

(f) the variety of substances that an organism drinks in a fixed period of time is increased (Dole, Ho, Gentry, & Chin, 1988),

(g) the substance flow during a drinking act is increased (Kardong & Haverly, 1993), and a variety of other intended meanings.

Each of the different possible *Increased drinking behavior* phenotypes may be the result of different underlying physiological causes, and it is therefore important to distinguish among them. A principal distinction regarding these phenotypes is between characteristics of single drinking processes and characteristics of processes with some duration in which drinking processes occur. Depending on the assay that is being used, only some of the qualities can be measured, while some may be inferred. For example, when the frequency of drinking processes that occur within a time period is *decreased*, and the total amount of liquid consumed is *increased*, then the liquid that is consumed in individual drinking acts must also be *increased* (on average, for each individual act of drinking).

NBO allows for the expression of distinctions between phenotypes of single process occurrences and multiple process occurrences. Therefore, we can distinguish between cases (a), (c), (e), (f) (which are phenotypes of multiple process occurrences) and (b), (d), (g) (which are phenotypes of single process occurrences). Using the PATO qualities, we can further make the type of process characteristic explicit. For example, we can use the *Increased frequency* (PATO:0000380) class in PATO to formalize case (e).

3. APPLICATION OF NBO

3.1. Human behavior phenotypes

Dissecting the genetic basis of behavior variation in humans is an important factor toward our understanding of human disease. The potential to identify the molecular underpinnings of human behavior and its characteristics depends on our ability to make meaningful genotype–phenotype correlations. Behavioral manifestations recorded in the clinic are not only an invaluable diagnostic tool but also provide insights to human pathophysiology and pathobiology. For example, the distinct behavioral characteristics of syndromes with known molecular basis such as the Angelman syndrome (hyperactivity, paroxysmal bursts of laughter, abnormal sleep patterns, ataxia) and Prader Willi syndrome (obsessive–compulsive features, learning difficulties, and language impairments) can help us understand the relations between genes and behavioral manifestations (Cassidy & Morris, 2002).

One useful resource that collects such information is the Online Mendelian Inheritance in Man (OMIM) database (Amberger, Bocchini, & Hamosh, 2011). OMIM presents a resource of signs and symptoms of human genetic resources as well as information about their genetic background when known. The Human Phenotype Ontology (HPO) (Robinson et al., 2008) provides annotations for a subset of OMIM entries. Previously, we have reported on our efforts of providing PATO-based logical definitions for HPO terms (Gkoutos et al., 2009). We have adopted the same approach and utilized NBO to describe behavior-related HPO terms. For example, the HPO term *Disinhibition* (HP:0000734) could be defined by combing the NBO term *social inhibition* (NBO:0000604) is linked to the *decreased rate* (PATO:00000911) term from the PATO ontology.

3.2. Mouse behavior phenotypes

Mouse presents one of the most important animal models for the study of behavior. There are numerous mouse models for the study of various aspects of behaviors such as anxiety (Finn, Rutledge-Gorman, & Crabbe, 2003), autism (Moy & Nadler, 2007), Parkinson's disease (Fleming, Fernagut, & Chesselet,

2005), DiGeorge Syndrome (Long et al., 2006), and Alzheimer's disease (Codita, Winblad, & Mohammed, 2006). The Mouse Genome Database (Bult et al., 2004; Part 2 Vol 104) serves as the model organism database for mouse and collects a variety of genetics and genomics related mouse information including mouse-related models and associated phenotypes. For the annotation of these phenotypes, it employs the Mammalian Phenotype (MP) Ontology (Smith, Goldsmith, & Eppig, 2004). We used PATO and NBO to formally decompose the MP classes that describe behavioral manifestations and thereby enable the integration of mouse behavior phenotype annotations with phenotype annotations from other species. For example, in order to formally define the MP term *decreased aggression toward mice* (MP:0003863), the NBO term *aggressive behavior toward mice* (NBO:0000107) is linked to the *decreased rate* (PATO:00000911) term from the PATO ontology.

3.3. Zebrafish behavior phenotypes

Zebrafish constitutes another invaluable animal model for human disease and has been employed for the study of complex neurological functions that affect behavior (Lieschke & Currie, 2007). A number of zebrafish behavior-related aspects are currently being tested including learning and memory, learning and cognition, conditioning, habituation, anxiety and aggression (Levin & Cerutti, 2009). The Zebrafish Model Organism Database (ZFIN) captures phenotype annotations from the literature originating from the zebrafish research community (Bradford et al., 2011). ZFIN curators annotate phenotype information following the PATO approach by combining the zebrafish anatomy ontology (http://zfin.org/zf_info/anatomy/dict/sum.html), GO and PATO. ZFIN currently contains 501 behavior-related phenotype annotations that have been created using GO behavior terms. Many of these annotations map to higher-level terms and lack the specificity that could be accomplished by utilizing NBO. ZFIN curators are currently in the process of working toward integrating NBO into their curation interface. This will allow back-curation and update of legacy behavior phenotypes but, more importantly, will be very beneficial for future curation with the influx of behavior phenotypes ZFIN expects with the large-scale mutagenesis screens that are in the pipeline.

3.4. *Drosophila* behavior phenotypes

Geneticists have been using *Drosophila* as a model genetic organism since the early 1900s. Fly models exist for the study of molecular mechanism of a wide range of human diseases, including neurodegenerative diseases. *Drosophila* behavior is a domain that is being thoroughly screened via a variety of behavioral assays that test a range of behavioral aspects including learning and memory, mating behavior, feeding behavior, circadian behavior, etc. (Nichols, Becnel, & Pandey, 2012). FlyBase is a community-driven model organism database that contains, among other type of data, phenotype information manually curated from *Drosophila* literature (Drysdale & FlyBase Consortium, 2008).

For the description of *Drosophila* phenotypes, FlyBase curators have adopted a controlled vocabulary of precomposed terms (FBcv) (Drysdale, 2001). We used PATO and NBO to formally decompose all the behavior-related phenotype classes that FBcv contains. For example, to define the FBcv term *chemosensitive behavior defective* (FBcv:000040), we combine the NBO term *chemosensory behavior* (NBO:0000322) with the PATO term *abnormal* (PATO:00000460) term.

3.5. Rat behavior phenotypes

Rats have been used as alternative model to mice for human cardiovascular disease, diabetes, arthritis, and many autoimmune and behavioral disorders. Rat behavior is a phenotypic aspect routinely assayed for various potentially genotype to phenotype and disease correlations (Cenci, Whishaw, & Schallert, 2002; Deumens, Blokland, & Prickaerts, 2002; Gilby, 2008; Liu & Wang, 2012). Based on their physiological and pathological similarity to humans, they are particularly useful for studying toxicity and pharmacodynamics of novel drugs. The Rat Genome Database (RGD, Volume 104, Part 2) forms a repository of rat genomic and genetic data, and RGD curators utilize a variety of different ontology for annotating biological information and have recently switched from the MeSH-based behavior vocabulary to NBO (Laulederkind et al., 2011). This change not only permits RGD-curated behavioral phenotypes to interlink with other biomedical ontologies, but more importantly, it will also facilitate the integration of rat behavioral observations within and across species.

4. DISCUSSION

The NBO is one of the first comprehensive ontologies designed for the integration of behavioral observations in animal organisms and humans. NBO's prime application is to provide the vocabulary that is required to integrate behavior observations within and across species. It is currently being applied by several model organism communities as well as for the description of human behavior-related disease phenotypes, and the use of a common, shared vocabulary for data annotation will lead to the possibility of integrative bioinformatics analyses of behavior-related data.

NBO also maintains compatibility with a wide variety of phenotype ontologies as well as with methods for postcomposing phenotypes at annotation time. To achieve these goals, NBO employs the PATO framework (Gkoutos, Green, Mallon, Hancock, & Davidson, 2005) of describing phenotypes a widely applied approach for formally characterizing phenotypes in multiple model organism databases as well as in the description of human disease phenotypes. The application of PATO for defining NBO classes leads to interoperability with these ontologies and their associated resources.

In addition to species-specific phenotype ontologies, several other efforts aim to provide ontologies that overlap with the behavior domain. For example, the GALEN ontology (Rector, Nowlan, & Glowinski, 1993) and SNOMED CT (Wang et al., 2001) provide comprehensive sets of clinical terms, some of which relate to behavior, and the emotion ontology (Hastings, Ceusters, Smith, & Mulligan, 2011) (for more information, see Chapter 5) specifically focus on terms that are relevant for describing emotions and moods. While the majority of these ontologies focus on human behavior and human behavioral phenotypes, it is an important area of future research to integrate other behavior-related ontologies with NBO. To achieve this goal, we may use lexical methods to establish mappings between other ontologies and NBO, and collaborate with ontology developers to align NBO with ontologies of other domains.

4.1. Relating animal models to human behavior-related diseases

Relating behavior-related processes in human and other animals is a challenging task for at least three main reasons. One of them relates to the conceptual and sometimes historical differences between clinical and lab approaches to describing behavior. The next refers to the potentially subtle differences between the actual behavior exhibited in a particular lab experiment and the subjective interpretation or correlation of the observations relating this experiment to human behavior (Gkoutos, Green, Mallon, Hancock, & Davidson, 2004c). Finally, there is an intrinsic genetic variation in normal and pathobiology between species (Schofield, Sundberg, Hoehndorf, & Gkoutos, 2011b). Undeniably though, animal models of human behavioral disorders are extremely valuable and their study has proven to be a powerful approach to our understanding of both human disease and fundamental mammalian biology. If we are to fully exploit the usefulness of animal models, it is imperative that we facilitate the integration of the large amounts of data that are being generated based on forward and reverse genetics, as well as pan-genomic phenotyping efforts (e.g., the International Mouse Phenotyping Consortium; Abbott, 2010). The NBO approach described here has been designed with the intention of serving that goal for the behavior-related aspect of those efforts. It is now included in two phenotype-based gene prioritization tools, PhenomeNet (Hoehndorf, Schofield, et al., 2011) and MouseFinder (Chen et al., 2012), and has proven successful in dissecting hereditary behavior diseases recorded in OMIM and OrphaNet. The NBO is one of the first ontologies exclusively dedicated to the annotation of behavioral phenotypes and is already widely applied across model organism communities and in bioinformatics projects. Its level of detail and specificity exceeds the information currently contained in species-specific phenotype ontologies, and therefore provides a valuable tool for research in behavioral neuroscience.

5. METHODS

5.1. Ontology

The initial version of the ontology was developed using a combination of OBO-edit (Richter, Harris, Haendel, & Lewis, 2007) and emacs. Subsequently, we transformed the ontology into the OWL format and it is currently maintained using Protege4 (Noy et al., 2001). In addition to simple relationships connecting classes, NBO contains a wide range of additional logical axioms, which are intended primarily assist with automated maintenance, quality control, and classification of the ontology.

5.2. NBO and phenotype ontologies

Phenotype ontologies usually contain descriptions of behavior-related manifestations. We have provided logical definitions based on NBO and PATO for three phenotypes ontologies, namely, MP, HPO, and FBcv. The relevant terms for each of these ontology was manually extracted and we subsequently provided equivalence axioms. For example, for the MP term *hyperdipsia* (MP:0005111), we provide the following computational definition:

```
'participates in' some
((regulates some 'drinking behavior')
and (has_quality some
('increased frequency'
```

```
and (towards some 'drinking behavior')
and (owl:qualifier some 'temporally extended'))))
```

We follow a similar procedure for defining the behavioral phenotype branch of NBO. For example, in order to define the NBO term *increased amount of liquid in a single drinking act* (NBO:0000851), we create the following definition:

```
'participates in' some
((has-input some
('liquid substance'
and (has_quality some 'increased mass')))
and (regulates some 'drinking behavior'))
```

5.3. Manual curation

The ontology was created via a combination of manual curation and computational reasoning. It was refined and populated via a combination of literature information, existing species-specific annotations, examination of behavior-related assays, personal communications with experts as well as our own domain knowledge. We also took into consideration a variety of existing ontologies that have behavior-related information such as MP and GO. We provide textual definitions for the NBO terms and where possible we provide links to their sources. We periodically realign the ontology with the existing phenotype ontologies by examining the change logs for different ontologies which we then manually check against NBO.

5.4. Maintenance, release, and availability

NBO is housed in a subversion repository and is made available via OBO registry and our project's Web site http://code.google.com/p/behavior-ontology/. There is a term request tracker http://code.google.com/p/behavior-ontology/issues/list and a discussion list https://lists.sourceforge.net/lists/listinfo/obo-behavior. NBO exists in two versions—an editor's version and a main release file. We make these versions available in OWL format and we utilize the OBO Ontology Release Tool (Oort) to converting the release versions into the OBO format, which we make available from our project.

ACKNOWLEDGMENTS

This work was supported by the National Institutes of Health (Grant number R01 HG004838-02) and the European Commission's 7th Framework Programme, RICORDO project (Grant number 248502).

REFERENCES

Abbott, A. (2010). Mouse project to find each gene's role. *Nature, 465*(7297).

Adcock, J. (2001). *Animal behavior: An evolutionary approach*. Sunderland, Massachusetts: Sinauer.

Amberger, J., Bocchini, C., & Hamosh, A. (2011). A new face and new challenges for online Mendelian inheritance in man (OMIM). *Human Mutation, 32*, 564–567.

Ashburner, M., Ball, C. A., Blake, J. A., Botstein, D., Butler, H., Cherry, M. J., et al. (2000). Gene ontology: Tool for the unification of biology. *Nature Genetics, 25*(1), 25–29.

Bada, M., Stevens, R., Goble, C., Gil, Y., Ashburner, M., Blake, J. A., et al. (2004). A short study on the success of the gene ontology. *Web Semantics: Science, Services and Agents on the World Wide Web, 1*(2), 235–240.

Barion, A. (2011). Circadian rhythm sleep disorders. *Disease-a-Month, 57*(8), 423–437.

Bolhuis, J., & Giraldeau, L. (2009). *Animal behaviour*. Thousand Oaks, California: SAGE: SAGE Library of Cognitive and Experimental Psychology.

Bradford, Y., Conlin, T., Dunn, N., Fashena, D., Frazer, K., Howe, D. G., et al. (2011). ZFIN: Enhancements and updates to the zebrafish model organism database. *Nucleic Acids Research, 39*(Database issue), D822–D829.

Brown, S. D., Chambon, P., de Angelis, M. H., & EumorphiaConsortium, (2005). EMPReSS: Standardized phenotype screens for functional annotation of the mouse genome. *Nature Genetics, 37*(11), 1155.

Bult, C. J., Blake, J. A., Richardson, J. E., Kadin, J. A., Eppig, J. T., Baldarelli, R. M., et al. (2004). The mouse genome database (MGD): Integrating biology with the genome. *Nucleic Acids Research, 32*(Database issue), D476–D481.

Cassidy, S. B., & Morris, C. A. (2002). Behavioral phenotypes in genetic syndromes: Genetic clues to human behavior. *Advances in Pediatrics, 49*, 59–86.

Cenci, M. A., Whishaw, I. Q., & Schallert, T. (2002). Animal models of neurological deficits: How relevant is the rat? *Nature Reviews. Neuroscience, 3*(7), 574–579.

Chen, C.-K., Mungall, C. J., Gkoutos, G. V., Doelken, S. C., Köhler, S., Ruef, B. J., et al. (2012). Mousefinder: Candidate disease genes from mouse phenotype data. *Human Mutation, 33*(5), 858–866.

Codita, A., Winblad, B., & Mohammed, A. H. (2006). Of mice and men: More neurobiology in dementia. *Current Opinion in Psychiatry, 19*(6), 555–563.

Congdon, E., Poldrack, R. A., & Freimer, N. B. (2010). Neurocognitive phenotypes and genetic dissection of disorders of brain and behavior. *Neuron, 68*(2), 218–230.

Deumens, R., Blokland, A., & Prickaerts, J. (2002). Modeling Parkinson's disease in rats: An evaluation of 6-ohda lesions of the nigrostriatal pathway. *Experimental Neurology, 175*(2), 303–317.

Dole, V. P., Ho, A., Gentry, R. T., & Chin, A. (1988). Toward an analogue of alcoholism in mice: Analysis of nongenetic variance in consumption of alcohol. *Proceedings of the National Academy of Sciences of the United States of America, 85*(3), 827–830.

Drysdale, R. (2001). Phenotypic data in FlyBase. *Briefings in Bioinformatics, 2*(1), 68–80.

Drysdale, R., & FlyBase Consortium, (2008). FlyBase: A database for the drosophila research community. *Methods in Molecular Biology (Clifton, N.J.), 420*, 45–59.

Finn, D. A., Rutledge-Gorman, M. T., & Crabbe, J. C. (2003). Genetic animal models of anxiety. *Neurogenetics, 4*(3), 109–135.

Fleming, S. M., Fernagut, P.-O., & Chesselet, M.-F. (2005). Genetic mouse models of Parkinsonism: Strengths and limitations. *NeuroRx: the Journal of the American Society for Experimental NeuroTherapeutics, 2*(3), 495–503.

Gilby, K. L. (2008). A new rat model for vulnerability to epilepsy and autism spectrum disorders. *Epilepsia, 49*(Suppl. 8), 108–110.

Gkoutos, G. V., Green, E., Mallon, A.-M., Hancock, J., & Davidson, D. (2004a). Using ontologies to describe mouse phenotypes. *Genome Biology, R8*.

Gkoutos, G. V., Green, E. C., Mallon, A. M., Hancock, J. M., & Davidson, D. (2004b). Building mouse phenotype ontologies. *Pacific Symposium on Biocomputing*, 178–189.

Gkoutos, G. V., Green, E. C. J., Mallon, A. M., Hancock, J. M., & Davidson, D. (2004c). Building mouse phenotype ontologies. In: R. B. Altman, K. A. Dunker, L. Hunter, T. A. Jung & T. E. Klein (Eds.), *Proceedings of the 9th Pacific symposium on biocomputing (PSB 2004), Hawaii, USA, January 6–10* (pp. 178–189), London: World Scientific.

Gkoutos, G. V., Green, E. C., Mallon, A.-M., Hancock, J. M., & Davidson, D. (2005). Using ontologies to describe mouse phenotypes. *Genome Biology*, *6*(1), R8.

Gkoutos, G. V., Mungall, C., Dolken, S., Ashburner, M., Lewis, S., Hancock, J., et al. (2009). Entity/quality-based logical definitions for the human skeletal phenome using PATO. In: *Conference Proceedings: . . . Annual International Conference of the IEEE Engineering in Medicine and Biology Society* (pp. 7069–7072).

Gooderham, P. A., Gagnon, R. F., & Gill, K. (2004). Attenuation of the alcohol preference of c57bl/6 mice during chronic renal failure. *The Journal of Laboratory and Clinical Medicine*, *143*(5), 292–300.

Grau, B., Horrocks, I., Motik, B., Parsia, B., Patelschneider, P., & Sattler, U. (2008). OWL 2: The next step for OWL. *Web Semantics: Science, Services and Agents on the World Wide Web*, *6*(4), 309–322.

Grier, J., & Burk, T. (1992). *Biology of animal behavior.* Saint Louis, MO: Mosby-Year Book.

Hailman, J. (1977). *Optical signals: Animal communication and light.* Bloomington, Indiana, USA: Indiana University Press.

Hamer, D. (2002). GENETICS: Rethinking behavior genetics. *Science*, *298*(5591), 71–72.

Hastings, J., Ceusters, W., Smith, B., & Mulligan, K. (2011). The emotion ontology: Enabling interdisciplinary research in the affective sciences. In: *Proceedings of the 7th international and interdisciplinary conference on modeling and using context. CONTEXT'11* (pp. 119–123), Berlin, Heidelberg: Springer-Verlag.

Hoehndorf, R., Dumontier, M., & Gkoutos, G. V. (2012). Identifying aberrant pathways through integrated analysis of knowledge in pharmacogenomics. *Bioinformatics*, *28*(16), 2169–2175.

Hoehndorf, R., Dumontier, M., Oellrich, A., Rebholz-Schuhmann, D., Schofield, P. N., & Gkoutos, G. V. (2011). Interoperability between biomedical ontologies through relation expansion, upper-level ontologies and automatic reasoning. *PloS One*, *6*(7), e22006.

Hoehndorf, R., Oellrich, A., & Rebholz-Schuhmann, D. (2010). Interoperability between phenotype and anatomy ontologies. *Bioinformatics*, *26*(24), 3112–3118.

Hoehndorf, R., Schofield, P. N., & Gkoutos, G. V. (2011). Phenomenet: A whole-phenome approach to disease gene discovery. *Nucleic Acids Research*, *39*(18), e119.

Holden, C. (2001). Animal behavior. Single gene dictates ant society. *Science*, *294*(5546), 1434.

Horrocks, I. (March 2007). OBO flat file format syntax and semantics and mapping to OWL Web Ontology Language. Technical Report, University of Manchester, http://www.cs.man.ac.uk/~horrocks/obo/. Accessed date 18/09/12.

Kardong, K., & Haverly, J. (1993). Drinking by the common boa, boa constrictor. *Copeia*, *3*, 808–818.

Laulederkind, S. J. F., Shimoyama, M., Hayman, G. T., Lowry, T. F., Nigam, R., Petri, V., et al. (2011). The rat genome database curation tool suite: A set of optimized software tools enabling efficient acquisition, organization, and presentation of biological data. *Database (Oxford)*, bar002.

Levin, E. D., & Cerutti, D. T. (2009). Chapter 15: Behavioral neuroscience of zebrafish. In *Methods of behavior analysis in neuroscience* (pp. 293–311). Boca Raton, Florida: CRC press.

Lieschke, G. J., & Currie, P. D. (2007). Animal models of human disease: Zebrafish swim into view. *Nature Reviews. Genetics*, *8*(5), 353–367.

Liu, X., & Wang, M. (2012). Gastrodin improves learning behavior in a rat model of Alzheimer's disease induced by intra-hippocampal Aβ 1–40 injection. *Molecular Neurodegeneration*, 7(Suppl. 1), S15.

Long, J., Laporte, P., Merscher, S., Funke, B., Saint-Jore, B., Puech, A., et al. (2006). Behavior of mice with mutations in the conserved region deleted in velocardiofacial/DiGeorge syndrome. *Neurogenetics*, 7(4), 247–257.

Mackay, T. (2008). The genetic architecture of complex behaviors: Lessons from drosophila. *Genetica*, *136*, 295–302.

Moy, S. S., & Nadler, J. J. (2007). Advances in behavioral genetics: Mouse models of autism. *Molecular Psychiatry*, *13*(1), 4–26.

Mungall, C., Gkoutos, G., Smith, C., Haendel, M., Lewis, S., & Ashburner, M. (2010). Integrating phenotype ontologies across multiple species. *Genome Biology*, *11*(1), R2.

Mungall, C., Torniai, C., Gkoutos, G., Lewis, S., & Haendel, M. (2012). Uberon, an integrative multi-species anatomy ontology. *Genome Biology*, *13*(1), R5.

Nichols, C. D., Becnel, J., & Pandey, U. B. (2012). Methods to assay drosophila behavior. *Journal of Visualized Experiments*, *61*(61), e3795.

Noy, N. F., Sintek, M., Decker, S., Crubezy, M., Fergerson, R. W., & Musen, M. A. (2001). Creating semantic web contents with Protege-2000. *IEEE Intelligent Systems*, *16*(2), 60–71.

Rector, A. L., Nowlan, W. A., & Glowinski, A. (1993). Goals for concept representation in the GALEN project. *Proceedings of the Annual Symposium on Computer Applications in Medical Care*, *1993*, 414–418.

Richter, J. D., Harris, M. A. A., Haendel, M., & Lewis, S. (2007). Obo-edit—An ontology editor for biologists. *Bioinformatics*, *23*, 2198–2200.

Robinson, P. N., Koehler, S., Bauer, S., Seelow, D., Horn, D., & Mundlos, S. (2008). The human phenotype ontology: A tool for annotating and analyzing human hereditary disease. *American Journal of Human Genetics*, *83*(5), 610–615.

Rose, H., & Rose, S. (2011). The legacies of Francis Galton. *The Lancet*, *377*(9775), 1397.

Schofield, P. N., Sundberg, J. P., Hoehndorf, R., & Gkoutos, G. V. (2011a). New approaches to the representation and analysis of phenotype knowledge in human diseases and their animal models. *Briefings in Functional Genomics*, *10*(5), 258–265.

Schofield, P. N., Sundberg, J. P., Hoehndorf, R., & Gkoutos, G. V. (2011b). New approaches to the representation and analysis of phenotype knowledge in human diseases and their animal models. *Briefings in Functional Genomics*, *10*(5), 258–265.

Searle, J. R. (1997). *The construction of social reality*. New York, NY: Free Press.

Smith, C. L., Goldsmith, C.-A. W., & Eppig, J. T. (2004). The mammalian phenotype ontology as a tool for annotating, analyzing and comparing phenotypic information. *Genome Biology*, *6*(1), R7.

Sokolowski, M. B. (2001). Drosophila: Genetics meets behaviour. *Nature Reviews. Genetics*, *2* (11), 879–890.

Spuhler, J. (2009). *Genetic diversity and human behavior*. Piscataway, New Jersey: Aldine Transaction.

Starr, C., & Taggart, R. (1998). *Cell biology and genetics. Biology series* (Vol. 1). Stamford, Connecticut: Wadsworth.

Tecott, L. H., & Nestler, E. J. (2004). Neurobehavioral assessment in the information age. *Nature Neuroscience*, 7(5), 462–466.

Tinbergen, N. (1963). On aims and methods of ethology. *Zeitschrift für Tierpsychologie*, *20*, 410–433.

Wang, A. Y., Barrett, J. W., Bentley, T., Markwell, D., Price, C., Spackman, K. A., et al. (2001). Mapping between SNOMED RT and clinical terms version 3: A key component of the SNOMED CT development process. *Proceedings of the Annual Symposium on Computer Applications in Medical Care*, 741–745.

Wehner, J. M., Radcliffe, R. A., & Bowers, B. J. (2001). Quantitative genetics and mouse behavior. *Annual Review of Neuroscience*, *24*, 845–867.

Wood, N. I., Goodman, A. O. G., van der Burg, J. M. M., Gazeau, V., Brundin, A., Björkqvist, P., et al. (2008). Increased thirst and drinking in Huntington's disease and the r6/2 mouse. *Brain Research Bulletin*, *76*(1–2), 70–79.

CHAPTER FIVE

Ontologies for Human Behavior Analysis and Their Application to Clinical Data

Janna Hastings[*,†,1], **Stefan Schulz**[‡]
[*]Swiss Center for Affective Sciences, University of Geneva, Geneva, Switzerland
[†]Cheminformatics and Metabolism, European Bioinformatics Institute, Cambridge, UK
[‡]Institute for Medical Informatics, Statistics and Documentation, Medical University of Graz, Graz, Austria
[1]Corresponding author: e-mail address: hastings@ebi.ac.uk

Contents

Abstract

Mental and behavioral disorders are common in all countries and represent a significant portion of the public health burden in developed nations. The human cost of these disorders is immense, yet treatment options for sufferers are currently limited, with many patients failing to respond sufficiently to currently available interventions.

Standardized terminologies facilitate data annotation and exchange for patient care, epidemiological analyses, and primary research into novel therapeutics. Such medical terminologies include SNOMED CT and ICD, which we describe here. Medical informatics is increasingly moving toward the adoption of formal ontologies, as they describe the nature of entities in reality and the relationships between them in such a fashion that they can be used for sophisticated automated reasoning and inference applications. An added benefit is that ontologies can be applied across different contexts in which traditionally separate domain-specific vocabularies have been used.

International Review of Neurobiology, Volume 103
ISSN 0074-7742
http://dx.doi.org/10.1016/B978-0-12-388408-4.00005-8

In this chapter, we report on a suite of ontologies currently in development for the description of human behavior, mental functioning, and mental disorders, and discuss their application in clinical contexts. We focus on the benefits of ontologies for clinical data management and for facilitating translational research for the development of novel therapeutics to treat challenging and debilitating conditions.

1. INTRODUCTION

Human behavior is one of the main indicators available to physicians to assess and infer underlying diseases and conditions, and monitor responses to treatments. It is especially relevant in the diagnosis and treatment of behavioral, psychological, and psychiatric conditions—such as obsessive–compulsive disorder, bipolar disorder, and schizophrenia, which we will jointly refer to hereafter as mental disorders—since in these conditions, there may be no other clinical indicators available.

Mental disorders are common in all countries, representing a significant portion of the public health burden. In the United States, about one in four adults is diagnosed with a mental disorder each year, and about one in 17 is thought to suffer from a serious and disabling mental illness (National Advisory Mental Health Council Workgroup, 2010). Mental disorders are the leading cause of disability in the United States and Canada for persons aged 15–44. The human cost of these disorders is immense, affecting not only patients but also their caregivers, rendering adults unable to work productively, destroying relationships, and increasing the financial burden on society. Treatment options for sufferers are currently limited, with many patients failing to respond sufficiently to currently available interventions, which include psychotherapeutic, somatic, and pharmacological actions. While there is enormous variance in individual responses to therapeutic agents, there is often little alternative for the clinician other than trial and error in determining the best treatment strategy given the patient's genetic, physiological, or behavioral profile.

Progress in primary research in many relevant frontiers of science is generating data that may be of relevance to address these challenges. Computer-based methods are essential to harness this ever growing body of data, information and knowledge, both in patient records and in scientific literature. Clinical decision-making processes in the treatment of individual patients need computational support, as do researchers in the interpretation of scientific findings. Traditionally, most relevant information has been available only as free and unstructured text. Machine processing, in contrast, necessitates adherence to terminological standards. This has led to ongoing community effort being invested in the development of standardized domain-specific terminology systems (Freitas, Schulz, & Moraes, 2009), for example, in the development of controlled vocabularies such as SNOMED CT (International Health Terminology Standards Development Organization, 2012) and classification systems such as the International Classification of Diseases (ICD) (World Health Organization, 2012b).

Ontologies are theories about the nature of entities in reality and the relationships between them. They are expressed in a logical language and enhanced with standard identifiers, labels, and definitions that facilitate unambiguous interpretation and annotation. A key advantage that ontologies confer over and above the mere standardization of terminologies is that their underlying logical formalisms are human language-independent and formally rigorous. This allows ontologies to form the backbone of sophisticated automated reasoning and inference applications. It also allows ontologies to be applied across contexts in which traditionally different domain-specific vocabularies have been used (Stenzhorn, Schulz, Boeker & Smith, 2008), facilitating interdisciplinary translation and disambiguation. Such ontologies are becoming widely used for the standardization and indexing of data in biological and medical domains (Munn & Smith, 2009; Rubin, Shah, & Noy, 2008; Smith, 2008).

In this chapter, we report on a suite of ontologies currently in development for the description of human behavior, mental functioning, and mental disorders, and discuss their application in clinical contexts. First, we focus on the benefits of ontologies for clinical data management and for facilitating translational research for the development of novel therapeutics to treat challenging and debilitating conditions. Second, we describe clinical vocabularies and terminologies that cover human behavior and mental functioning. Third, we describe ontologies that are currently under development to formalize the entities and relationships underlying human functioning and disorder. Finally, our discussion and conclusion focus on the varying applications of these ontologies in clinical data management and in translational research.

2. MEDICAL TERMINOLOGIES AND VOCABULARIES FOR HUMAN FUNCTIONING

2.1. SNOMED CT

The Systematized Nomenclature of Medicine Clinical Terms (SNOMED CT; International Health Terminology Standards Development Organization, 2012) is a multi-hierarchical medical terminology system that provides codes, terms, synonyms, and definitions. It aims to provide a consistent way to index, store, retrieve, and aggregate clinical data across disciplines and locations. It contains 311,000 representational units, called SNOMED CT concepts, that cover all aspects of the Electronic Health Record (EHR). At present, it is organized along 19 different semantic axes or subhierarchies, including "clinical finding," "body structure," "observable entity," "disorder," and "organisms." Clinical findings are the elements of a diagnosis and are often related to a particular "observable entity." In the case of human behavior, for example, there is an observable entity called "Behavior observable" which has classification parent "Mental state, behavior/psychosocial function observable" and classification children "Ability to control behavior," "Aspect of behavior," "Behavior of childhood and adolescence," "Behavioral assessment of the dysexecutive syndrome score," "Behavioral phenotype," "Characteristic of complex/social behavior," "Habits," "Health-related behavior," "Interpretation of behavior," "Motor function behavior," "Personal autonomy behavior," "Predictability of behavior," "Safe wandering behavior of cognitively impaired subject," and "Safety behavior." For these observable entities, related clinical findings (linked to the observable entity with an "interprets" relation) include "Manic behavior" and "Withdrawn behavior." In total, 963 SNOMED CT preferred terms contain the string "behavior*." Most of these (713) are in the finding hierarchy, of which 509 are classified as disorders (as a type of finding). In all, 132 are in the observable entity hierarchy and 64 in the procedure hierarchy.

2.2. ICD and ICF

While SNOMED is focused on the standardization of clinical data, the World Health Organization (WHO) is maintaining a multilingual vocabulary to classify diseases, ICD (World Health Organization, 2012b), currently in its 10th version, with the 11th in preparation. The ICD is intended as a standard classification for diseases in epidemiological and clinical applications. In particular, ICD was originally created to assist with the statistical task of monitoring the incidence and prevalence of diseases across countries, populations, and specific subgroups, linked to variables such as socioeconomic status. ICD annotations provide the basis for national mortality and morbidity statistics as reported by the WHO for member states. ICD 10's chapter V is dedicated to "Mental and behavioral disorders." Examples of classes included in chapter V are "behavioral and emotional disorders with onset usually occurring in childhood and adolescence," "Schizophrenia, schizotypal, and delusional disorders," "Mood (affective) disorders," and

"Unspecified mental disorder." Note that the use of an "unspecified" catch-all category is common in medical terminologies as a holder category for annotation of disorders that do not fit within any of the alternative categories on the same level. The string "behavior*" occurs in 245 titles. However, this includes some in which behavior is used to refer to the malignancy of tumors, for example, "Neoplasm of uncertain or unknown behavior of oral cavity and digestive organs." As a peculiarity of the ICD chapter V, its codes are provided by detailed glossary entries, which are absent in the remaining chapters of ICD.

A related project to ICD within the WHO standardization effort is the International Classification of Functioning, Disability and Health (ICF) (World Health Organization, 2012a). The ICF provides a classification of health and health-related domains, including bodily functions and structure, domains of activity and participation, and environmental factors. Within the classification of body functions, there is a chapter devoted to mental functions such as consciousness, energy and drive, memory, language, and calculation. Within the classification of activities and participation, there are chapters devoted to learning and applying knowledge, communication, interpersonal interactions and relationships, and community and social life. Within the classification of environmental factors, there are chapters devoted to support and relationships as well as attitudes. The ICF is intended to provide description of the factors in the ordinary functioning of humans, dysfunctioning of which may be implicated in a variety of the disorders and conditions listed in ICD. However, specific "behavior" categories are missing in the ICF.

ICD is, by far, the most important WHO terminology. Used around the globe, it is available in all major languages and thus constitutes one of very few universal terminology standards. However, ICF and SNOMED CT coded data are still restricted to selected health institutions, mainly in the United States, the UK, and the Northern countries. WHO and IHTSDO, however, have signed a cooperation agreement, one manifestation of which is the joint development of an ontological core of future ICD versions.

2.3. DSM-IV

In the domain of psychiatry, a related classification system called the Diagnostic and Statistical Manual for Mental Disorders (DSM) (APA, 2000) is widely used for the classification of diagnoses of mental disorders of relevance. While DSM was engineered to refer to ICD codes, this is only partially in place due to the asynchronous evolution of both systems. Unlike the ICD, the DSM provides not only a classification of disorders but also guidance as to the diagnostic criteria for these disorders in the form of checklists of symptoms, with counts of how many symptoms of each sort are needed for a positive diagnosis. The DSM is currently in its fourth revision, but the fifth revision is scheduled for release in May 2013 (Regier, Narrow, Kuhl, & Kupfer, 2009), and a draft version of the revisions have been released for public review at www.dsm5.org. Some issues that the revision will try to address are a high occurrence of co-morbidity of disorders according to the diagnostic criteria and the high use of "catch-all" categories such as "not otherwise specified." To address these, the revision is expected to emphasize dimensional measures of symptoms that cross diagnostic category boundaries.

3. FROM CLINICAL TERMINOLOGIES TO ONTOLOGIES

3.1. Domain and upper-level ontologies

Modern biomedical ontologies offer several advantages over clinical terminologies and this fact is contributing to the success and uptake of these ontologies. In contrast to terminology systems, which focus on representing and standardizing the meaning of terms as units of human language, ontologies aim to provide a formal account of objects of the world independently of language. They can be (and usually are) enhanced by the annotation of labels and definitions in a human-understandable language that guides the interpretation and application of the ontology for various purposes. Many existing systems in clinical contexts are actually hybrids, with some terminological and some ontological aspects. For example, in the terminology SNOMED CT the meaning of terms is partially supported by logical axioms, and in the ICD classification, terms and their natural language definitions are designed to assign the objects of interest (patients and/or their diseases) into disjoint categories.

Domain ontologies, such as the Gene Ontology (The Gene Ontology Consortium, 2000) and ChEBI (Chemical Entities of Biological Interest; de Matos et al., 2010), are increasingly rooted in upper ontologies such as the Basic Formal Ontology (BFO) (Grenon & Smith, 2004; Smith, 2012), DOLCE (Gangemi, Guarino, Masolo, Oltramari, & Schneider, 2002), GFO (Herre et al., 2006), and BioTop (Beisswanger, Schulz, Stenzhorn, & Hahn, 2008). Upper level ontologies provide a basic set of (mostly) mutually disjoint categories, enriched by constraining logical axioms that allow computers to check for errors and ensure consistency. Alignment of a domain ontology to an upper-level ontology involves the selection of the most appropriate upper-level category for each entity in the domain ontology. Ontologies based on the methodology of ontological realism (Smith & Ceusters, 2010) focus on the accurate description of the portion of reality covered by the ontology, which necessitates clearly distinguishing between information entities, such as a diagnosis, that can be mistaken, and the disease that the patient actually suffers from. As we will see in what follows, such core ontological distinctions are of paramount importance in the annotation of clinical data for the analysis of human behavior.

3.2. Mental functioning ontology

Based on the upper-level ontology BFO and being developed in the context of the OBO Foundry (Smith et al., 2007) library of interrelated modular domain ontologies, the Mental Functioning Ontology (MF) (Hastings, Ceusters, Jensen, Mulligan & Smith, 2012) is a new overarching modular domain ontology covering all aspects of mental functioning, including mental processes such as cognition, and traits such as intelligence. MF provides a solid grounding for mental functioning entities in an upper-level ontology and gives a framework within which mental functioning can be related to alternate levels of description within other ontologies, such as anatomy and biochemistry. Modules that are currently actively under development are those for cognition, perception, and emotion.

Figure 5.1 illustrates the upper levels of the ontology, based on the framework laid out in Ceusters & Smith (2010a), together with the alignment to BFO. At the top level, BFO introduces a distinction between continuants and occurrents. While occurrents are, in a rough sense, processes and other entities that unfold in time, continuants are those things that continue to exist over an extended period of time, such as organisms. This distinction can be seen in the context of mental functioning between, for example, a part of an organism's *brain anatomy* that continues to exist over time thus is a continuant, and an organism's *thinking process* that spans over a few minutes and is then completed. Within continuants, BFO further distinguishes between those that are independent and those that are dependent. Independent continuants can exist by themselves, while dependent continuants are those sorts of things that need a "bearer" in order to exist, such as colors, or dispositions that are realized in patterns of behavior.

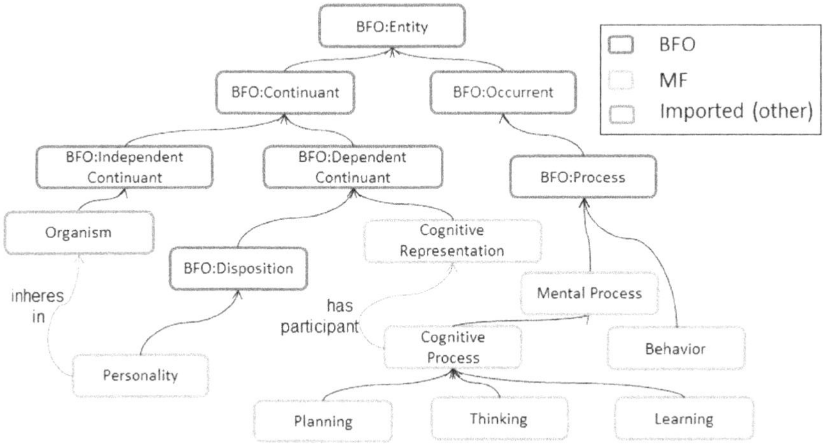

Figure 5.1 Mental Functioning Ontology upper-level alignment to BFO: Mental functions are capabilities that inhere in organisms such as human beings. These functions are realized in mental processes, such as planning, thinking, remembering, or undergoing an emotion. Functioning takes place by virtue of underlying physiological, biochemical, and anatomical configurations, which are classified as mental functioning related anatomical structures. Personality is a disposition that inheres in a person and is realized in the (characteristic) behavior of that person. (For interpretation of the references to color in this figure legend, the reader is referred to the online version of this chapter.)

The illustrated upper levels of MF show several important distinctions in the framework to annotate and describe mental functioning allowing interrelationships across a wide variety of different levels of description. The organism is the fundamental independent continuant in which mental functioning takes place. A mental functioning related anatomical structure is that part of an organism that bears a disposition to be the agent of a particular mental process. So, for example, the particular neuronal and biochemical configuration (i.e., the *bona fide* group of receptors and neurotransmitters (Ceusters & Smith, 2010b)) that gives rise to a particular person's feeling of sadness is a mental functioning related anatomical structure. Neurons and brain chemistry are themselves described as continuants in other ontologies such as ChEBI for the neurotransmitters, the Protein Ontology (Natale et al., 2011) for the receptors, and NeuroLex and BIRNlex (Bug et al., 2008) for neurons and neuronal systems. These components can be linked together as parts of the corresponding mental functioning related anatomical structure, the boundaries of which are to be determined with the advance of our understanding of the neurobiology and neurochemistry of the physical basis of the various mental processes involved. The links from entities in MF to the known biochemical and neurobiological bases will be maintained in bridging modules, ensuring that different levels of granularity and description can be separately maintained. References to other vocabularies such as ICD and BIRNlex will be annotated in the ontology where applicable.

Dispositions are properties that inhere in their bearers by virtue of what will happen when the bearer comes into the right circumstances, for example, a glass breaking when it is dropped onto a hard surface. An example of a disposition in the domain of mental functioning is human personality, since personality (or character) is the sort of property that is realized in the behavioral interactions of the human being with the external world, as well as in patterns of thought or performance in learning new things and other mental processes. Personality may be measured by standardized tests (which are information content entities) in application scenarios in multiple domains including psychology and human resources. These tests can ideally be linked using a "measures" relation to the personality attributes in MF.

As for cognitive or mental occurrents, MF includes mental processes, which are defined as the processes that manipulate, bring into being, or destroy cognitive representations. Cognitive representations are themselves dependent continuants that specifically depend on the cognitive structures of an organism and represent cognitive content that can take the form of thoughts or memories. Mental processes—manipulating those cognitive representations—include all of the standard processual examples of mental functioning such as thinking, planning, learning, or remembering.

MF is being developed modularly, allowing different teams with different core areas of expertise to focus on the extension of the overall ontology to describe the entities relevant to their scientific area. One such extension is the Emotion Ontology (MFO-EM; Hastings, Ceusters, Smith & Mulligan, 2011), describing entities of relevance to all aspects of affective science. Figure 5.2 illustrates the extension of MF for emotion-related entities.

Emotional action tendencies are dispositions to behavior that arise from emotions, for example, the characteristic "fight or flight" action tendency that arises from fear. The physiological response to an emotion involves physiological changes in the central nervous system and neuro-endocrine system. An emotional behavioral process is a behavior that straightforwardly results from the emotion, such as facial expression changes in response to the emotion. Characteristic facial expression changes are considered by many to be intrinsic parts of the unfolding of emotion. For this reason, pictures of characteristic facial expressions are a predominant paradigm in cognitive (affective) neuroscience: subjects are shown emotional faces, while they are undergoing

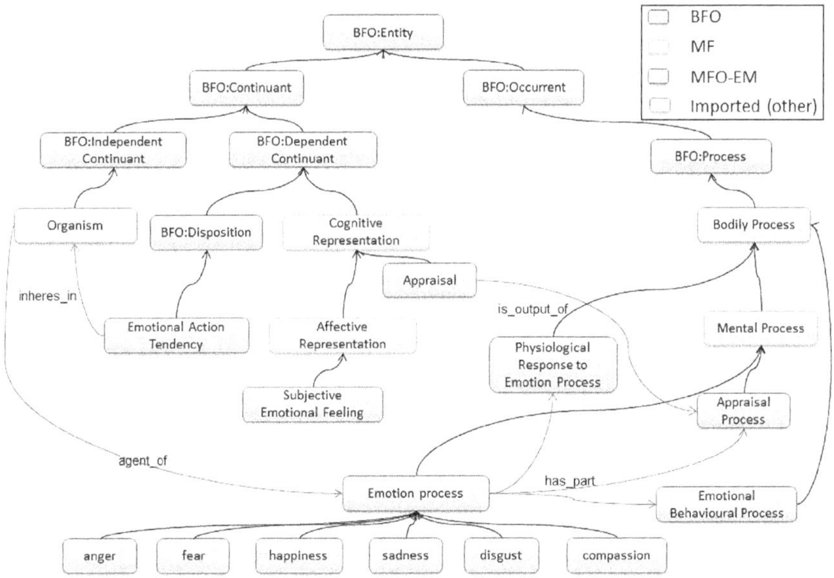

Figure 5.2 Emotion Ontology upper-levels beneath MF: Emotions are complex synchronized processes with physiological and mental components. The components include a physiological response (such as sweating), behavior (such as an expression of shock), a subjective feeling (such as a sense of inner coldness), and an action tendency (such as the urge to run away). Each component has been classified in the EM ontology, as illustrated. (For interpretation of the references to color in this figure legend, the reader is referred to the online version of this chapter.)

functional neuroimaging experiments. Such paradigms for cognitive neuroimaging are being described in the Cognitive Paradigm Ontology (Turner & Laird, 2012), with which the Emotion Ontology is currently being aligned. Interlinking emotion research in different communities—such as that of cognitive neuroimaging and, for example, genetics, or model organisms—is facilitated by shared annotation to ontologies that represent what the research is about, rather than just the standard terminologies used in each community. The Emotion Ontology is currently being applied in two separate application scenarios: first, in the capture of self-reported emotions, and second, in the meta-analysis of brain imaging results across multiple diverse studies.

3.3. Mental disease ontology

Another MF extension covers the domain of mental disease. A critical ontological question that is not fully addressed by the DSM, ICD, or SNOMED is the question of what a mental disorder actually *is*. The Mental

Disease Ontology (MD) is a separate ontology module that aims to describe and categorize mental disorders based on the core definitions and extension strategy outlined in Ceusters and Smith (2010a). The MD extends not only the MF but also the Ontology for General Medical Science (OGMS). OGMS is designed to interrelate ontologies in the medical domain to support research on EHR technology and on the integration of clinical and research data and provides definitions for disease, diagnosis, and disorder based on the terminology outlined in Scheuermann, Ceusters, and Smith (2009).

Following OGMS, a mental disease is defined as a disposition to undergo pathological mental processes. A mental disease is a clinically significant deviation from mental health. Mental health is conformity of perception, emotion, and behavior internally and in relation to the external real-world environment. In contrast, pathological mental processes are those that hinder well-being. Thus, mental disease is a deviation from mental health that hampers the bearer in his or her mental well-being (Ceusters & Smith, 2010a). Figure 5.3 shows an extract of entities from MD for the domain of substance addiction, a mental disease characterized by substance use and phenomena such as tolerance, craving, and withdrawal. Figure 5.3 shows an extract of entities from MD for the domain of substance addiction, a mental disease characterized by substance use and phenomena such as tolerance, craving, and withdrawal symptoms.

For each mental disease, the ontology contains representations of the symptoms and signs that are manifested in the disease course, including pathological behavior. By differentiating a disease from a disease course and by explicitly representing symptoms and signs within a logically rigorous ontological framework that includes a definition for mental disease, MF aims to address some of the challenges that have been observed with the DSM approach, such as high levels of comorbidity and the use of catch-all "not otherwise specified" placements. The DSM approach, termed "descriptive psychiatry," focuses on symptom assessment and confers disorder status on specified thresholds of symptoms in terms of counts of symptom types and tokens and durations of symptom episodes. For example, a major depressive episode is stated to be diagnosable if five of a set of nine symptoms are found to obtain within the same 2-week period. Symptoms include "insomnia or hypersomnia nearly every day" and "fatigue or loss of energy nearly every day." (Notice how these are not likely to be mutually exclusive.) The DSM-5 proposal has also been criticized for promoting medicalization of *normal human experiences*: grief, a normal human emotion in response to bereavement, has been proposed as a type of depression, a mental disorder (Cacciatore, 2012).

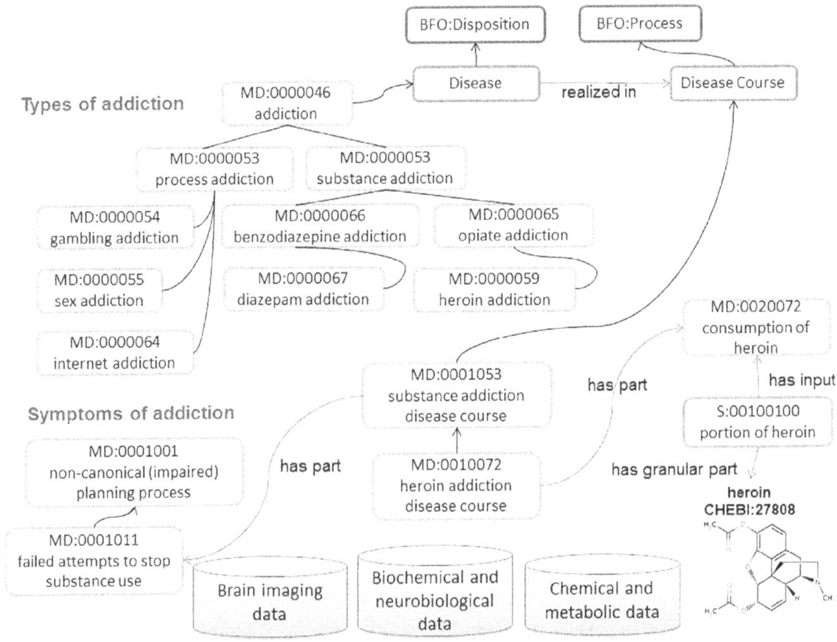

Figure 5.3 Addiction in MD: The MD follows OGMS in distinguishing between diseases as dispositions, and the disease courses in which disease dispositions are realized as pathological processes. In the case of addiction, the disease hierarchy distinguishes many different types of addiction based on the object of the addiction, which also correspond to distinctions in the underlying pathophysiological pathways. Disease courses contain symptoms as parts, for example, the substance addiction disease course contains repeated failed attempts to stop substance use, a kind of pathological planning process, as a part. The heroin addiction disease course contains consumption of heroin as a part. We illustrate the interlinking of biologically relevant knowledge that is obtained via bridging modules between bio-ontologies: consumption of heroin is linked via the portion of substance that is consumed to the description of heroin in the ChEBI ontology, and thereby to related chemical and metabolic knowledge bases. (For interpretation of the references to color in this figure legend, the reader is referred to the online version of this chapter.)

One symptom of substance addiction, for example, is a preoccupation with use of the substance in question, a kind of noncanonical (i.e., not in accordance with the environment, not conducive to well-being) thinking process (because the organism is not able to control the thinking process as they would in canonical thinking processes). Furthermore, pathological (or noncanonical) processes are related to the canonical versions of those processes. This interlinking of symptoms to diseases and to canonical related processes in a computable framework allows bridging from research involving different diseases to research exploring ordinary functioning or underlying mechanisms. It also allows hypotheses of mechanisms underlying diseases to be made explicitly testable in terms of supporting data.

3.4. Ontologies in the analysis of human behavior

While it is easy to intuitively understand what is meant by "human behavior," the precise definition proves quite elusive on closer examination. While many would agree, for example, that eating is a behavior, and that chewing is a behavior, there would be more disagreement about whether digestion and other autonomous processes are behaviors. We can ask ourselves what sort of thing is human behavior as follows. From the perspective of ontological types, MF has classified behavior as a process, meaning that it is always an occurrent that unfolds in time, rather than continuing to exist through time. But it is not always straightforward to find the correct ontological category, even for rather generic terms like behavior. An organism's behavior can be interpreted as a process which is presently going on, for example, the mating behavior of a fruit fly, classified as a process in the Gene Ontology. On the other hand, some might also interpret behavior as a disposition inhering in an organism even when it is not currently being manifested, which explains the fact that in SNOMED CT many behavior concepts are classified under "observable quality"; in MF dispositions of this sort would be either action tendencies or personality attributes.

It is also difficult to analyze how behavior is distinct from other processes that take place in human organisms. Autonomous processes such as metabolizing, digesting, circulation, and breathing are not traditionally considered as behaviors. Neither are processes that occur only once for an organism, such as birth or death. Whereas digestion is not itself a behavior, we might nevertheless include "eating behavior" as a type of behavior and be interested in certain characteristic patterns or attributes that further describe the process of eating by an organism.

A complication in efforts to analyze mental functioning is that behavior is the only readily observable aspect of mental functioning—unlike thoughts, feelings, and other aspects of mental functioning. Indeed, while brain imaging and other technological advances may develop to the extent where correlates of certain sorts of mental functioning are reliably measurable, there is as yet no evidence that this is guaranteed to be possible and as yet no such technology exists. Mental functioning—and related mental disorders—is thus a particularly challenging area of medicine, and consequently for the annotation of clinical data. Technology that allows description of behavior and related mental functioning is thus an essential tool for the annotation of data contributing to this research effort. Annotation of characteristic behavior in different model organisms is also an essential component in cross-species research paradigms in biology and medicine, since in those cases nothing resembling patient reports or clinical interviews are accessible.

Ontologies such as we have described above contain unambiguous definitions for what is intended to be implied by their entities labeled with terminology such as "behavior" that can, in ordinary discourse, be interpreted in many different ways. One of the ways that this has relevance in the clinic is in the design and delivery of clinical questionnaires and diagnostic tools. Annotating such data as results from such diagnostic tools in a consistent fashion by multiple different investigators, physicians, and patients in such a fashion so as to ensure that consistent results are obtained from application of the instrument is an area of great importance for the definitional clarifications offered by modern ontologies.

4. APPLICATIONS TO CLINICAL DATA AND TRANSLATIONAL RESEARCH

As modern clinical contexts become increasingly computerized, management of clinical data and ease of use by medical practitioners become increasing priorities. Of core importance in this context is the ontological distinction between, on the one hand, the clinical information model, which includes facts such as that the gender of a particular patient or the diagnosis of another may be unknown, and on the other hand, the patients and diseases themselves that certainly do have specific genders and specific disease types. Ontological annotation is also essential in maximizing the benefit of clinical data, such as in the EHR system of a hospital or medical facility, for purposes such as reporting and clinical research.

One aspect of clinical data management is that of the organization and maintenance of biobank data in which human samples are stored for purposes of clinical research (Krestyaninova, Spjuth, Hastings, Dietrich, & Rebholz-Schuhmann, 2011). In order to research underlying mechanistic factors in rare diseases, samples from patients bearing the condition may need to be sourced from multiple biobanks in multiple countries or regions. Traditional systems that use local terminologies (language and country-specific) terminologies to annotate the sample databases will certainly not be straightforward to integrate and search across different sample collections. It is even more difficult to interrelate sample data with EHR data and with known indicators in medical and biological knowledge-bases such as those collecting annotated genetic sequence information. The need for shared ontologies to annotate these diverse clinical data is becoming widely recognized.

There are many areas of medicine where mental functioning has unexpected influence on medical treatment for other conditions. One well-known factor is that of psychological effects, such as the placebo effect, the nocebo effect, and the treatment effect. The placebo effect is that in which taking a treatment that is merely believed to be beneficial but has no actual active component can result in positive effects. The nocebo effect is the opposite: negative consequences produced by an inert treatment, based on negative expectation. The treatment effect is a very interesting and well-known effect of relevance particularly in clinical contexts, namely, that offering *some* treatment for a given condition produces an experience of recovery, usually attributed to the treatment by the patient, even in cases where the treatment had no causal role to play in the recovery of the patient. These effects are so standard that they need to be factored into all research in clinical contexts and drug discovery. Formalizing the description of such phenomena in ontologies allows the annotation of research into their neural and biochemical correlates.

Genetic and psychiatric population-wide research often relies on diagnostic interviews which standardize the collection of data into aspects of psychiatric functioning such that the data can be compared and aggregated across large groups of patients. In the domain of mental functioning, this is a particularly pressing problem since many aspects of mental functioning are not directly observable, and the assessment of mental functioning therefore relies on the subjective assessment of the trained practitioner and on self-reports by the patient, who of course has no access to alternative experiences of mental functioning other than his/her own. Standardized questionnaires are thus an essential element of population research into mental functioning. An example of such a questionnaire is the Diagnostic Interview for Genetic Studies (Nurnberger et al., 1994), a questionnaire used in clinical interviews to assess major mood and psychotic disorders and related spectrum conditions. Linking the symptoms assessed in such questionnaires to ontologies of mental functioning provides the capability to standardize the collected data across multiple such questionnaires. Furthermore, it allows multilevel aggregation, rather than only aggregation at the level of whether a particular disorder is diagnosed or not—which in some cases may obscure rather than illuminate shared underlying mechanisms and pathologies.

Increasing the speed and throughput of the translation of primary research in brain and mind science into novel therapeutic agents, and ultimately clinical interventions, has been highlighted as a pressing current concern for mental health research and practice (National Advisory Mental Health Council Workgroup, 2010). However, this effort is hindered by the disconnect between the different communities involved in primary research and the different levels needed for the translation into therapeutics. Understanding the processes involved in mental disorders requires research and integration of knowledge across all the different levels of life science, from the most fundamental such as genetic and biomolecular, through medical, brain, and neurosciences, to the psychological and psychiatric perspectives which focus on the behavioral and functional aspects. Recent breakthroughs in basic science in all of these different levels have the potential to be exploited toward novel interventions and therapeutics, but severe obstacles remain in the path of translation, and there is still a resulting shortage of new agents and approaches in the therapeutic pipeline (National Advisory Mental Health Council Workgroup, 2010). Most importantly, ontologies offer a common language that enables automated bridging between different disciplines, facilitating translation as research becomes increasingly interdisciplinary. Furthermore, sophisticated querying and hypothesis testing frameworks are able to be developed around the ontologies.

Ontology provision does not bring about these translational benefits single-handedly. Complementary efforts are needed to bring about a simultaneous revolution in data sharing and community practices to enable all relevant data to be annotated with ontologies, integrated, and thereby made available for the entire research community. One such effort is the Neuroscience Information Framework (NIF), which provides a powerful ontology-backed portal for searching and discovery across all databases and other data resources of relevance for neuroscience (Gardner et al., 2008). NIF incorporates ontologies such as MF that are being developed by the community and assists in the efforts of semantically interlinking different ontologies through bridging modules. Another contemporary effort is the One Mind for Research project (1mind4research.org), which gathers and indexes data resources and acts within the community to promote a culture of sharing for translational research.

5. CONCLUSIONS

Ontologies are becoming increasingly important throughout many modern clinical and biomedical contexts, from patient interactions in the form of structured questionnaires and physician reporting, to translational research for the development of novel treatments for challenging conditions. Human behavioral analysis is a challenging topic that bridges across several related ontology projects and annotation needs. We have surveyed the behavior-related content of widely used medical terminologies such as SNOMED and ICD and described ongoing work in community-based ontology development projects for mental functioning, disease and emotion. As scientific knowledge across a growing number of different levels of description of biomedical reality is accumulated in disparate domain-specific ontology projects, a pressing challenge becomes the scientifically relevant interrelationships between those ontologies to allow automated bridging between the domains and to facilitate translational data-driven research.

ACKNOWLEDGMENTS

We thank Colin Batchelor, Jane Lomax, David Osumi-Sutherland and George Gkoutos for discussions on the topic of behavior. We further wish to thank all contributors to the Mental Functioning Ontology project, particularly Werner Ceusters, Mark Jensen, and Barry Smith. J. H. thanks the EU for funding under the OPENSCREEN project, work package "Standardization." The content of this chapter is solely the responsibility of the authors.

REFERENCES

APA, (2000). *Diagnostic and Statistical Manual of Mental Disorders, Fourth Edition—Text Revision*. Washington, DC: American Psychiatric Association.

Beisswanger, E., Schulz, S., Stenzhorn, H., & Hahn, U. (2008). BioTop: An upper domain ontology for the life sciences—A description of its current structure, contents, and interfaces to obo ontologies. *Applied Ontology*, *3*, 205–212.

Bug, W., Ascoli, G., Grethe, J., Gupta, A., Fennema-Notestine, C., Laird, A., et al. (2008). The NIFSTD and BIRNLex vocabularies: Building comprehensive ontologies for neuroscience. *Neuroinformatics*, *6*(3), 175–194.

Cacciatore, J. (2012). DSM5 and ethical relativism. http://drjoanne.blogspot.com/2012/03/relativity-applies-to-physics-not.html. Accessed April 2012.

Ceusters, W., & Smith, B. (2010a). Foundations for a realist ontology of mental disease. *Journal of Biomedical Semantics*, *1*(1), 10.

Ceusters, W., & Smith, B. (2010b). A unified framework for biomedical terminologies and ontologies. *Studies in Health Technology and Informatics*, *160*, 1050–1054.

de Matos, P., Alcántara, R., Dekker, A., Ennis, M., Hastings, J., Haug, K., et al. (2010). Chemical Entities of Biological Interest: An update. *Nucleic Acids Research*, *38*, D249–D254.

Freitas, F., Schulz, S., & Moraes, E. (2009). Survey of current terminologies and ontologies in biology and medicine. *RECIIS—Electronic Journal in Communication, Information and Innovation in Health*, *3*(1), 7–18.

Gangemi, A., Guarino, N., Masolo, C., Oltramari, A., & Schneider, L. (2002). Sweetening ontologies with DOLCE. In: *Proceedings of EKAW 2002* (pp. 166–181), Berlin, Heidelberg: Springer. Vol. 2473 of LNCS.

Gardner, D., Akil, H., Ascoli, G. A., Bowden, D. M., Bug, W., Donohue, D. E., et al. (2008). The Neuroscience Information Framework: A data and knowledge environment for neuroscience. *Neuroinformatics*, *6*(3), 149–160.

Grenon, P., & Smith, B. (2004). SNAP and SPAN: Towards dynamic spatial ontology. *Spatial Cognition & Computation: An Interdisciplinary Journal, 4*(1), 69–104.

Hastings, J., Ceusters, W., Jensen, M., Mulligan, K., & Smith, B. (2012). Representing mental functioning: Ontologies for mental health and disease. In: *ICBO 2012 Workshop, Towards an Ontology of Mental Functioning. Graz, Austria; July 22, 2012.*

Hastings, J., Ceusters, W., Smith, B., & Mulligan, K. (2011). Dispositions and processes in the Emotion Ontology. In: *Proceedings of the International Conference on Biomedical Ontology (ICBO2011), Buffalo, USA.*

Herre, H., Heller, B., Burek, P., Hoehndorf, R., Loebe, F., & Michalek, H. (2006). General Formal Ontology (GFO)–A Foundational Ontology Integrating Objects and Processes [Version 1.0]. Technical Report 8, Research Group Ontologies in Medicine, Institute of Medical Informatics, Statistics and Epidemiology, University of Leipzig, Leipzig.

International Health Terminology Standards Development Organization. (2012). Systematized nomenclature of medicine—Clinical terms (SNOMED-CT). http://www.ihtsdo.org/snomed-ct/. Accessed May 2012.

Krestyaninova, M., Spjuth, O., Hastings, J., Dietrich, J., & Rebholz-Schuhmann, D. (2011). Biobank metaportal to enhance collaborative research: Sail.simbioms.org. In: *Proceedings of ICTA 2011, Orlando, Florida.*

Munn, K., & Smith, B. (Eds.), (February 2009). *Applied ontology: An introduction.* Ontos Verlag.

Natale, D. A., Arighi, C. N., Barker, W. C., Blake, J. A., Bult, C. J., Caudy, M., et al. (2011). The Protein Ontology: A structured representation of protein forms and complexes. *Nucleic Acids Research, 39* (Database issue), D539–D545.

National Advisory Mental Health Council Workgroup. (2010). From discovery to cure: Accelerating the development of new and personalized interventions for mental illness. http://www.nimh.nih.gov/about/advisory-boards-and-groups/namhc/reports/fromdiscoverytocure.pdf. Accessed October 2012.

Nurnberger, J. I., Jr., Blehar, M. C., Kaufmann, C. A., York-Cooler, C., Simpson, S. G., Harkavy-Friedman, J., et al. (1994). Diagnostic interview for genetic studies: Rationale, unique features, and training. *Archives of General Psychiatry, 51*(11), 849–859.

Regier, D. A., Narrow, W. E., Kuhl, E. A., & Kupfer, D. J. (2009). The conceptual development of DSM-V. *The American Journal of Psychiatry, 166*, 645–650.

Rubin, D. L., Shah, N. H., & Noy, N. F. (2008). Biomedical ontologies: A functional perspective. *Briefings in Bioinformatics, 9*(1), 75–90.

Scheuermann, R., Ceusters, W., & Smith, B. (2009). Toward an ontological treatment of disease and diagnosis. In: *AMIA Summit on Translational Bioinformatics, San Francisco, California, March 15-17, 2009* (pp. 116–120), Omnipress.

Smith, B. (2008). *Ontology (science).* In: Proceedings of the 2008 conference on Formal Ontology in Information Systems: Proceedings of the Fifth International Conference (FOIS 2008) (pp. 21–35), Amsterdam, The Netherlands: IOS Press. http://dl.acm.org/citation.cfm?id=1563953.1563958. Accessed October 2012.

Smith, B. (2012). BFO 2.0 Draft. http://ontology.buffalo.edu/bfo/Reference/. Accessed January 2012.

Smith, B., Ashburner, M., Rosse, C., Bard, J., Bug, W., Ceusters, W., et al. (2007). The OBO Foundry: Coordinated evolution of ontologies to support biomedical data integration. *Nature Biotechnology, 25*(11), 1251–1255.

Smith, B., & Ceusters, W. (2010). Ontological realism as a methodology for coordinated evolution of scientific ontologies. *Applied Ontology, 5*, 139–188.

Stenzhorn, H., Schulz, S., Boeker, M., & Smith, B. (2008). Adapting clinical ontologies in real-world environments. *Journal of Universal Computer Science, 14*(22), 3767–3780.

The Gene Ontology Consortium, (2000). Gene ontology: Tool for the unification of biology. *Nature Genetics*, *25*, 25–29.
Turner, J. A., & Laird, A. R. (2012). The cognitive paradigm ontology: Design and application. *Neuroinformatics*, *10*(1), 57–66.
World Health Organization. (2012a). International classification of functioning, disability and health (ICF). http://www.who.int/classifications/icf. Accessed March 2012.
World Health Organization. (2012b). International statistical classification of diseases (ICD). http://www.who.int/classifications/icd. Accessed March 2012.

CHAPTER SIX

Text-Mining and Neuroscience

Kyle H. Ambert[1], Aaron M. Cohen
Department of Medical Informatics & Clinical Epidemiology, Oregon Health & Science University, Portland, OR, USA
[1]Corresponding author: e-mail address: ambertk@ohsu.edu

Contents

Abstract

The wealth and diversity of neuroscience research are inherent characteristics of the discipline that can give rise to some complications. As the field continues to expand, we generate a great deal of data about all aspects, and from multiple perspectives, of the brain, its chemistry, biology, and how these affect behavior. The vast majority of research scientists cannot afford to spend their time combing the literature to find every article related to their research, nor do they wish to spend time adjusting their neuroanatomical vocabulary to communicate with other subdomains in the neurosciences. As such, there has been a recent increase in the amount of informatics research devoted to developing digital resources for neuroscience research. Neuroinformatics is concerned with the development of computational tools to further our understanding of the brain and to make sense of the vast amount of information that neuroscientists generate (French & Pavlidis, 2007). Many of these tools are related to the use of textual data. Here, we review some of the recent developments for better using the vast amount of textual information generated in neuroscience research and publication and suggest several use cases that will demonstrate how bench neuroscientists can take advantage of the resources that are available.

International Review of Neurobiology, Volume 103
ISSN 0074-7742
http://dx.doi.org/10.1016/B978-0-12-388408-4.00006-X

1. INTRODUCTION

Like most domains in biological research, neuroscience has experienced a recent explosion in the volume of published information (Shepherd et al., 1998). The history of neuroscience can arguably be traced back at least as far as the works of Camillo Golgi and Santiago Ramón y Cajal, in the early twentieth century. Since that time, neuroscience has become increasingly fractionated into various subdomains, incorporating elements of molecular biology, genetics, computer science, and cognitive science, to name but a handful. Each of these domains has proven equally prolific, such that a simple Google Scholar search for "neuro*" yields nearly a million and a half results. To say that any one scientist can or should have this volume of information available for immediate recall in his or her head is folly, and yet, in order to efficiently advance the field of research, this can seem exactly what would be required. How can we, as neuroscientists, be sure we are not repeating ourselves, investigating experimental hypotheses that have long since been addressed? How can scientists efficiently synthesize the knowledge within a particular neuroscientific subdomain in order to see where the gaps in our knowledge lie? Given the diversity of training background in the neuroscience community, how can we be sure we are not falling subject to communication errors, using differing terminology to refer to similar neuroanatomical concepts, and therefore losing opportunities to make new conceptual connections? These are the kinds of questions that neuroinformatics and text-mining attempt to address. Each of these questions has been posed in the past, and a variety of solutions have been devised. Several of the solutions that have shown to provide greatest benefit, and most potential for continued use, are derived from a subdomain of machine learning called text-mining. In this chapter, we review many of the important developments in text-mining research as well as how they apply and can be applied to research the behavioral neurosciences.

2. TERMINOLOGIES AND DATA INTEGRATION

Neuroscience is an incredibly diverse field, consisting of researchers from many disciplines. Although united by a shared interest in the study of the brain, each field has its own way of communicating—the cognitive psychologist might refer to Brodmann area IV, while the behavioral neuroscientist might refer to the primary motor cortex. Researchers in the field are not typically confused by this diversity in language, but computers often are. To the non-informatician, this may not seem like much of a problem—after all, computers do not need to "understand" concepts, they just need to efficiently manipulate them in accordance with a user's instructions. Unfortunately, this is very much not the case. Although neuroinformatics is still a young field, the heterogeneity of terms in neuroscience is already an interesting problem being addressed in order to improve mathematical modeling, machine learning document classification systems, and information retrieval (IR) systems, with a particular focus on neuroanatomical terminologies. Terminologies can be helpful tools for facilitating communication between colleagues in related disciplines and subdisciplines and aid in data sharing. Ontologies are related, as they allow for the definition of hierarchical types of objects and abstract concepts in a way that is understandable to both machines and human readers. Here we discuss two example systems: NeuroNames, and the NIFSTD and BIRNLex Ontologies.

3. NEURONAMES

Co-created by Douglas Bowden and Richard Martin (Bowden & Martin, 1995; Martin, Dubach, & Bowden, 1990), NeuroNames (http://braininfo.rprc.washington.edu/) was one of the first popular neuroanatomical terminologies in the field. At the time it was first published, there was an absence of machine-readable neuroanatomical terminologies, making even something as seemingly straightforward as finding articles pertaining to a particular neuroscience subdiscipline difficult (Bowden & Dubach, 2003). In order to facilitate scholarly communication and IR in the neurosciences, Bowden and colleagues set out to define a "comprehensive set of mutually exclusive primary structures that constitute the brain" (Bowden & Dubach, 2003). NeuroNames consists of 15,000 neuroanatomical terms, spanning 2500 brain-related concepts, culled from textbooks, atlases, and research articles (Bowden, Dubach, & Park, 2007). One of the most important contributions of the NeuroNames vocabulary is that it constitutes one of the first attempts to standardize neuroanatomical terms, by serving as a reference point for neuroscientists and by providing a standardized set of terms that unites multiply-defined anatomical structures by combining the concept name and the author and year of publication of the publication in which the term appeared (e.g., area 9 of Brodmann-1909).

4. LEVERAGING NEUROSCIENCE ONTOLOGIES AND VOCABULARIES

The Neuroscience Information Framework (NIF) has made significant contributions to fulfilling the need for standardized terminologies in neurosciences. Their standardized ontology (the NIFSTD) is an hierarchically structured collection of neuroscience-related terminologies, including terms used for describing neuroscience data, methods, anatomy, and digital resources (Bug et al., 2008; Imam et al., 2012). The project is an extension of the Biomedical Informatics Research Network project (Martone, Gupta, & Ellisman, 2004), is formatted in the style of a semantic wiki, as the NeuroLex, (Bandrowski, 2011; Larson, Iman, Bakker, Pham, & Martone, 2010; Larson & Martone, 2009), and is easily downloadable in *OWL* file format http://purl.org/nif/ontology/nif.owl, the standard format for describing ontologies (see Fig. 6.1 for an example). The idea is that neuroinformaticians developing their own resources will be inclined to fold the NIFSTD ontology into their own resources, rather than developing a new set of terms, as has so often been the case in the past.

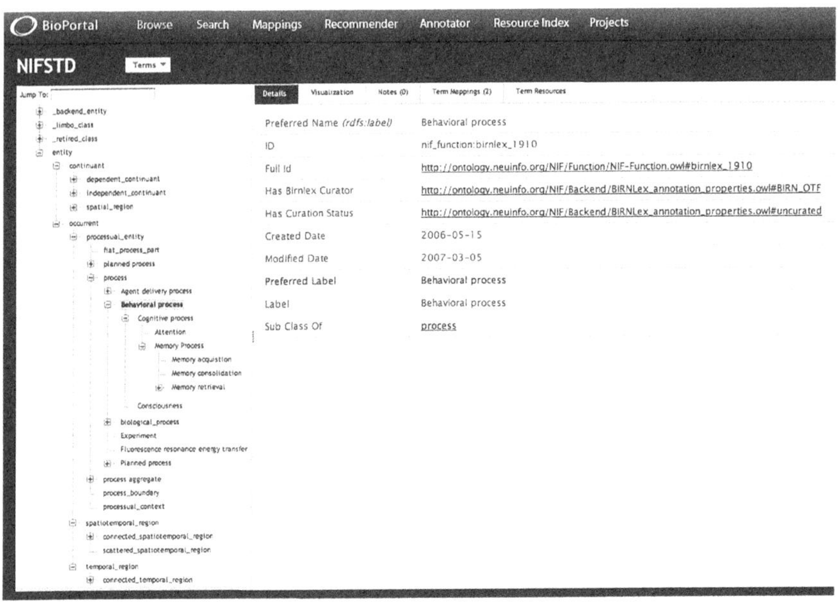

Figure 6.1 Screen shot of the NIFSTD ontology OWL format viewed using the BioPortal ontology viewer (http://bioportal.bioontology.org). (For color version of this figure, the reader is referred to the online version of this chapter.)

In fact, this movement has already begun to take hold. For example, Maynard, Mungall, Lewis, and Martone (2010) used the NIFSTD to connect entities in clinical descriptions of human disease to model systems, thus bridging phenotypes in animal models from behavioral research to descriptions of human pathological features.

On the surface, terminologies and ontologies may not seem like useful resources to bench neuroscientists, as they seem something far removed from their day-to-day research activities. However, they begin to address what has long been recognized as a difficult problem that is deeply integrated into the way neuroscientists think about the brain. Sometimes called the *neuron classification problem* (Bota & Swanson, 2007), the question of what constitutes necessary and sufficient criteria for distinguishing one type of neuron from another, dates back to the foundation of neuroscience itself, with Camillo Golgi and Santiago Ramón y Cajal (Clarke & Jacyna, 1987). Are histological differences sufficient for distinguishing one cell type from another, or should spatial location in the brain be a factor as well? Within a particular region of the brain (e.g., central nucleus of the amygdala), is directionality also important (e.g., lateral, ventral)? These are the questions that neuroinformaticians, in collaboration with molecular neuroanatomists, aim to address. The decisions that are made will facilitate how researchers interact with one another, both in terms of scholarly discourse (e.g., how we describe neuron-related findings), as well as in terms of how they share data with each another. As users, other neuroscientists will benefit from further development of these tools by being able to better collaborate with other researchers in related disciplines.

5. INFORMATION RETRIEVAL

IR is a subdiscipline of computer science that is concerned with developing accurate algorithms for retrieving information from databases of documents or textual information (Hersh, 2009). In general, IR systems are designed to take users' search requests (queries), identify relevant data in a database, and return a ranked list of results that is ordered according to likelihood of relevance to the input query (Hersh, 2009). Such systems are quite common in today's information-heavy age, with common examples being Google search, PubMed, or Apple's Spotlight system, on the OSX operating system. In the biomedical sciences, IR is most commonly associated with the National Library of Medicine's PubMed search engine (http://www.ncbi.nim.nih.gov/pubmed/), which queries against a database of over 21 million peer-reviewed scientific publications. In addition to joining query terms via standard Boolean operators (e.g., AND, NOT, OR; PubMed Help), PubMed also utilizes a vector representation of the query to identify the most relevant related articles (Jensen, Saric, & Bork, 2006). Although PubMed is one of the first resources many researchers will use when performing a literature search, it is not without its limitations.

Domain-specific IR systems can provide several advantages over general-purpose ones, such as PubMed. Although general-purpose biomedical IR solutions will often suffice, there are situations where neuroscientists can need specialized search tools (Ascoli, 2012). For example, a researcher conducting a literature review on retrograde tracer studies could run a simple PubMed query *retrograde tracer* and obtain approximately 2700 results (query performed on July, 2012). The enumerated publications will be articles in which the term *retrograde* and the term *tracer* appeared in at least 1 of 64 data fields (e.g., Abstract, MeSH Term, Title; for a full and up-to-date list, see http://www.nlm.nih.gov/bsd/mms/medlineelements.html). If the researcher is only interested in studies that actually used retrograde tracing as an experimental method, the results returned by PubMed are likely to contain many documents that are not of interest (e.g., 35 of the results obtained were review articles), and, in addition, are likely to not identify publications that would have been relevant (e.g., studies that used retrograde tracing, but did not include this exact term in the titles or abstracts). Because of this, the researcher performing the literature review will have to spend time manually going through the entire list of results to identify the publications that are genuine of interest, in addition to performing extra queries, to obtain articles that were not initially identified. The costs associated with performing these tasks are often prohibitive; thus, neuroinformaticians have constructed specialized search tools for the neuroscience literature base that can overcome this difficulty. Two of the major developments in neuroscience information retrieval (NIR) solutions that have come about in the last 5 years are Textpresso for Neuroscience (Müller, Rangarajan, Teal, & Sternberg, 2008) and the platform developed by the Neuroscience Information Framework (Gardner et al., 2008). As these two systems have taken somewhat different approaches to addressing NIR, we will discuss each in turn.

6. TEXTPRESSO FOR NEUROSCIENCE

Textpresso for Neuroscience is neuroscience-specific version of the popular Textpresso system, from Müller, Kenny, and Sternberg (2004), of Howard Hughes Medical Institute and California Institute of Technology.

Textpresso is an IR system distinguished by two key components: the ability to perform full-text searches and the use of ontology (see Section 4), allowing for defining types of objects and abstract concepts in a way that is understandable by both machines and human readers. One can easily perform a search for *anchor cell* in a general-purpose search engine, but many of the documents returned may end up being, for example, about maritime justice systems. If we are truly interested in documents only referring to anchor cells in the biological sense, ontology could be useful for informing the search system that anchor cells are a type of biological cell in the *C. elegans* and are characterized by production of the signaling molecule LIN-3/EGF (Hill & Sternberg, 1992). To allow full-text searching, Textpresso uses the xpdf software (http://www.fooiabs.com/xpdf/) in combination with journal-specific templates, which allow them to extract the plain text from the PDF representation of a publication with some degree of accuracy. This approach contrasts with that taken by PubMed, which uses publisher-supplied metadata (e.g., keywords) for their database. Although this approach is limited somewhat by the hit-or-miss process of extracting text from a PDF, it does allow users to query against the entire document, which can be advantageous, particularly if users wish to query based on text that is likely to be found in figure captions (Hirschman et al., 2012). Similarly advantageous is Textpresso's use of ontologies to facilitate accurate searching of the text. In the original Textpresso paper (Müller et al., 2004), Müller and colleagues describe a variety of categories that were used to mark up their documents, enabling a variety of concepts to be included in a search query, including biological concepts, relationships, and descriptions (Müller et al., 2008). For example, to search for brain areas in which the TRP channel TRPC1 is found, the user could specify to include TRPC1 and select the categories *brain area* and *NIF (neural) stem cell types.*

To extend their approach to neurosciences, Müller and colleagues included publications from 18 neuroscience journals that were selected in collaboration with the NIF (Müller et al., 2008). As of the time of this writing, their system allows full-text searching for over 100,000 neuroscience publications and allows for the specification of several neuroscience-related term categories and subcategories (Table 6.1). Textpresso for Neuroscience can be accessed either through the systems' main Web site (http://www.textpresso.org/neuroscience/), or through their Web service. In addition, it has been incorporated into the NIF (Gupta et al., 2008). The Textpresso for Neuroscience system can been used by research scientists outside of neuroinformatics to further their own work. Because the Textpresso system allows for full-text searching of research publications, users can perform more

Table 6.1 Neuroscience-specific categories, approximate size of their lexica (in terms of number of words and phrases), and example terms

Category	Number of terms in Lexicon	Example terms
Brain area	4800	Terminal sulcus, area 1 of Brodmann-1909
Drugs of abuse	190	Alcohol, heroin
Nicotine addiction (NICSNP) candidate gene	380	GIRK6, VAMP4
NIF cell type	138	Horizontal cells
Neuropsychology and behavior	125	Hebbian pairing, saccade
Prescription drug of abuse	105	Robitussin A-C, Ritalin
Receptor	5700	Metabotropic glutamate receptor 8
Substance abuse	73	Self-administration, addiction
TRP channel	40	TRPV1

Reproduced with permission from Müller et al. (2008).

specific queries that are targeted at text occurring throughout the document. If one is interested in retrieving documents based on information that is in figure captions (where experimental results are frequently described with greater concision), this would be possible with Textpresso, since the entire text is indexed, but it would only be possible for the open access publications that are indexed by PubMed. A major limitation of the system, however, is that its bibliography has not been updated since 2009 (*Web site accessed on July, 2012*). This highlights a shortcoming of many digital resources: it is typically more common for research scientists to receive grant funding for a project aiming to develop new methods for using or accessing digital resources than it is for one that will maintain said resource beyond its initial funding period. An incredibly useful tool, such as Textpresso for Neuroscience, is only as good as the data it indexes, and since the number of neuroscience-related publications is always increasing, without ongoing support it can quickly become out of date. It is our hope that this trend will change in the future. One resource, which we turn to now, has a great track record of maintaining its relevance—the Neuroscience Information Framework.

7. IR USING THE NEUROSCIENCE INFORMATION FRAMEWORK

The Neuroscience Information Framework was created as a part of the National Institutes of Health's Blueprint for Neuroscience Research in 2004 (Baughman, Farkas, Guzman, & Huerta, 2006; Gardner et al., 2008). A complete description of the NIF can be found in Chapter 3 of this volume. Briefly, the NIF distinguishes itself from more traditional document IR systems (e.g., PubMed) by providing a central framework with which existing online neuroscience resources can be integrated. These resources are not just limited to documents—they include expression data (e.g., as documented in BrainSpan http://www.brainspan.org/), atlases (e.g., as documented in the Allen Mouse Brain Atlas http://www.brain-map.org/), and imaging databases (e.g., as documented in the Brede Database http://neuro.imm.dtu.dk/services/jerne/brede/). This diversification stems from the NIF's driving goal, to facilitate access to, and integration of, heterogenous neuroscience data, for the purpose of enabling new discoveries to be made and new neuroinformatics tools to be developed (Gardner et al., 2008).

Integrating dynamically updated data from geographically distributed resources can be something of a daunting task, since all data needs to be mapped from different views of the human brain into a common data model, but, if carried out properly, it provides significant advantages to users. The NIF currently offers three levels of data integration to neuroscientists who have information resources they would like to make available. The most in-depth of these levels allows contributors to integrate their data into the larger NIF data federation by submitting schema information and database views to the NIF mediator. They use a concept mapping tool to map the data to the tables, fields, and values in the NeuroLex ontology (http://neurolex.org). This allows resource providers to leave their data in its original format, maintaining its integrity and leaving any necessary transformations to be made in the ontology mapping stage. This allows for updates to the content to be made available as they happen. From the perspective of the user, this deep-level integration means that queries performed on the NIF's main page will be run against a variety of neuroscience data resource simultaneously, with the results packaged in a way that is meaningful and easy to navigate. For example, running the query *Amygdala basolateral nucleus pyramidal neuron* on the NIF returns 189 literature results, and several results from the data federation—four brain regions, two genes, four grants, and two diseases (*query performed on July, 2012*). If more than one of these resource categories were of interest to a user, and he or she was not using the NIF, multiple queries would need to be performed on several external databases (e.g., BAMS, OMIM, and NIH RePORTER) using different query formats and terminologies, which would be time-consuming to perform, and would leave the scientist to do the integration of the retrieved results.

One use case for a resource like the NIF is that of data integration. Because the NIF takes care of mapping multiple heterogenous data resources back to common data ontology, it is possible to query across multiple data types in a meaningful way. To return to the *Amygdala basolateral nucleus pyramidal neuron* query example, if a scientist were interested in doing a study involving this cell type, he or she could learn that four grants have been funded to NIH institutions on this topic, but that the most recent one ended in 2011. One would also find that, in the Online Mendelian Inheritance in Man (OMIM) database, it related to brain-derived neurotrophic factor, obsessive-compulsive disorder, and congenital central hypoventilation syndrome. All of this information would be helpful to developing a new hypothesis or designing a study, and it is immediately available in one integrated resource.

A second use case relates more directly to text-mining experiments that might be conducted by or for behavioral neuroscientists. Behavioral assays, such as the elevated plus maze (Rodgers & Dalvi, 1997), conditioned place preference (Cunningham, Gremel, & Groblewski, 2006), or the adjusting-amount procedure (Mitchell & Rosenthal, 2003), are the backbone of behavioral neuroscience. Such procedures are used as behavioral models of disease and used, for example, to evaluate the efficacy of drugs for treating disease. If a scientist were conducting a literature review on the use of the adjusting-amount procedure in evaluating the effects of dopamine-2 receptor antagonists on impulsive choice, they could perform a query in PubMed, and manually shift through the many documents it would return. Carrying out the same task using the NIF, however, would allow the researcher to leverage the previously described ontology, ensuring that the results returned are indeed relevant to both the behavioral procedure in question and the specific class of drugs. That is, the results would include instances of the procedure and drug themselves, rather than just the words themselves (i.e., *adjusting-amount procedure* as a method, rather than documents containing the words *adjusting-amount* and *procedure*). As it stands, this tool is useful enough, but the future possibilities for this type of IR could greatly affect the way literature reviews are conducted in the behavioral sciences. For example, using a procedure similar to that described in the CoCoMac classification experiment described in section 9, one could use the NIF to obtain documents in which certain behavioral procedures are known to have been used. These data could be used to create a document classifier that would then identify research publications in which the procedure was used, but which had not been identified by the NIF either because they were newly published or because of publisher error.

8. SUPERVISED TEXT CLASSIFICATION

The frequency and volume of newly published scientific literature is quickly making the maintenance of publicly available scientific databases unrealistic and costly. Assuming a newly published article is identified as potentially containing relevant information, database curators can spend up to 48 h determining whether it should be included in their database, and manually extracting the relevant information from the full-text document. Therefore, supervised document classification systems are an increasingly effective machine learning tool to promote efficiency for the many text-related tasks in biomedical science (Cohen & Hersh, 2005). In such systems, a collection of documents are manually annotated with regards to some criteria—for example, include/exclude in a database, or relevant/irrelevant for a literature review, and are then used to train a classifier to make judgments on documents that have not yet been seen. Cohen and colleagues (Cohen, Adams, et al., 2010; Cohen, Ambert, & McDonagh, 2009; Yang, Cohen, McDonagh, 2008) have used such an approach to provide text-mining support tools to the systematic review community. In this work, the Medline records associated with documents are used as input features to a classifier that assigns each a relevance judgment for a number of systematic review topics. In a more biomedical application, they have also used text classification for using the text in the i2b2 challenge tasks for mining clinical discharge summaries to predict smoking status (Cohen, 2008), obesity-related disease comorbidity status (Ambert & Cohen, 2009), and identification of biomedical concepts, assertions, and relations (e.g., type II diabetes, "disease is present," and "hypertension was controlled by hydrochlorothiazide," respectively) (Ambert & Cohen, 2011; Cohen, Ambert, et al., 2010).

In the neurosciences, document classification is manifest in the maintenance of databases documenting primary source experimental data on, for example, neuroanatomical connectivity. Many of these databases have become invaluable resources for neuroscientists studying connectivity itself (Bohland et al., 2009; Sporns, Tononi, & Edelman, 2000) and a useful reference for behavioral neuroscientists in conducting lesion or microinjection studies. Despite the frequency with which they are used, the information contained in such connectivity databases is often based on user-submitted connection information, and it may not be possible for the database owner to find enough time to verify the information, or to identify new information to update the database.

Gully Burns and colleagues' Scientific Knowledge Mine (SciKnowMine) project is an important development for behavioral researchers (Helmer et al., 2011; Ramakrishnan et al., 2012). They recently showed how their document classification/biocuration pipeline can be used to help curation at the Mouse Genome Informatics group (Bult et al., 2008). They take an all-in-one approach to solving the problem of applied text-mining, providing a system that stores documents, extracts text from PDFs, preprocesses data, maps the text to ontology, and outputs the data to Web services. They used this system at the MGI to perform automated document triage (identifying which documents in a large data set are irrelevant for some curation task). Burns and colleagues' unified system approach to text-mining is an important example of how machine learning experts and neuroinformaticians are beginning to recognize the importance of making their tools accessible and useful for performing common tasks in research scientists' workflows; it will likely be a model for future text-mining system developments in the future. Similarly, the work of Lynette Hirschman, Gully Burns, and others (Burns, Feng, & Hovy, 2008; Burns, Krallinger, Cohen, Wu, & Hirschman, 2009; Hirschman et al., 2012; Pokkunuri, Ramakrishnan, Riloff, Hovy, & Burns, 2011) has shown how text-mining can be used to optimize biocuration workflows in the molecular sciences. In particular, text-mining can be useful for the document triage task described above, wherein bio-entity identification and normalization (i.e., removing specific mentions of biological entities from text prior to classification) can be leveraged to develop a useful document classification system or to suggest relations for annotation in a database. For example, in a recent study where we built a document classifier for identifying protein–protein interaction (PPI)-related information (Ambert & Cohen, 2011), we observed that replacing protein mentions in the text of documents with a normalized feature (e.g., changing "5-HT Receptor" to "PROTEIN_MENTION") led to improved classification performance. The reason for this is that in many biocuration classification procedures, it is more important that the classifier use the contextual features surrounding annotatable information than the specific entities themselves. In the case of neuroanatomical connection classification, this would be akin to relying more on features like *connects*, *afferent*, and *efferent*, rather than ones like *hippocampus*, *cortex*, and *striatum*. Similar to the PPI normalization case described above, the contextual features will allow the classifier to more easily identify documents containing annotatable information regarding neuroanatomy that it has not previously seen.

9. CLASSIFICATION FOR THE COCOMAC DATABASE—AN EXAMPLE OF TEXT-MINING FOR NEUROSCIENCES

Text classification experiments can be fairly complex, but as a rule of thumb, there are generally five elements to a text classification pipeline:

1. *Text extraction*: Free text is extracted from a PDF document (e.g., in Ramakrishnan et al., 2012), Web site, or some other input resource, and put in a format readable by the classification software. This could be a directory of *txt* files, an *xml* file, or a database.
2. *Pre-processing*: This step is important to get the extracted text into a regularized and predictable form (Ambert & Cohen, 2011). In the above-mentioned PPI study, we found that an important feature of a document classifier for identifying papers containing PPI-related information was a step in which we removed all mentions of specific proteins. Classification systems make their judgments based on the characteristics of the input documents. Thus, if one's goal is to create a system for identifying documents containing a variety of PPIs, and not just those that were observed in the training data, removing specific PPI mentions forces the classifier to make its judgments based on other document characteristics, for example, the sorts of sentence structures that often describe relation information between two proteins (e.g., "our data demonstrate that PROTEIN interacts with PROTEIN"). Other procedures frequently done during preprocessing are the removal of all punctuation in the text and case-normalization.
3. *Tokenization*: In this step, the preprocessed documents are split into individual tokens or features. A simple normalization procedure that is frequently used in text-mining experiments is simple unigram tokenization. This approach splits the document into a "bag of words,"

wherein each feature is a word and no ordering is conserved. Other approaches will be based on bi- or tri-grams (individual pairs or trios of words, respectively), which retain some word ordering observed in the original document.

4. *Modeling*: The collection of tokens resulting from the tokenization step is next modeled for use by the classification algorithm. Binary feature modeling is a commonly used modeling procedure in which the unique set of features observed in the entire training document collection is assigned a position in vector. Each document is then represented as vector of the same length, in which each position contains either a zero or a one, corresponding to the absence or presence of that feature within the document in question.
5. *Classification*: The classification algorithm is given a set of (*vector, true class label*) pairs (during classifier training) or just document vectors (during classification), and using whatever classification procedure has been selected for the task, it will either learn the mathematical relationship between document feature vectors and their class labels (in training), or predict the class label of new documents (during classification). Many classification algorithms exist, but Support Vector Machines (SVMs; Joachims, 1998) and Naïve Bayes (McCallum & Nigam, 1998) are commonly used procedures in text classification.

As a proof of concept for the application of text classification in neurosciences, we developed a machine learning framework for automating the identification of sentences containing neuroanatomical connectivity information appropriated for incorporation into the CoCoMac online database of Macaque connectivity information (http://www.CoCoMac.org). The CoCoMac database was selected for several reasons. First, it contains a great deal of connectivity information indexed according to the PubMed Identifier (PMID) associated with the article from which the information was obtained. Many online neuroscientific databases contain a combination of unpublished experimental data and peer-reviewed results, and since this proof-of-concept system is concerned with verifying the information that has been accepted into the scientific body of knowledge, it made sense to choose a database specifically focusing on the published literature. Second, the CoCoMac database has an intuitive, built-in URL search interface that makes it easy for an automated system to pull down information on an as-needed basis, rather than having one or more individuals spend time performing manual IR. Third, CoCoMac's article curation process is rigorous and well-documented. Furthermore, the CoCoMac database has not been updated since 2005, due, according to its founder, to the fact that verifying the information contained in one article can take up to 2 days (Rolf Kotter, 2009; personal communication)—emphasizing the need for automated methods for streamlining the curation process.

We created a classifier that, given a list of connections supposedly documented within an article, would identify the sentences in the article's abstract containing this information. Our general workflow for system development is diagramed in Fig. 6.2. We first obtained a complete list of PMID IDs contained in the CoCoMac database (approximately 600 IDs) and located an electronic version of the full text for each using PubMed, Google, and Google Scholar. Even though the present set of experiments was based on sentence-level classification judgments in the abstract, an important follow-up experiment is to expand our classification to Results sections in full text, as well, and therefore our studies included only those abstracts for which we could obtain the entire document (approximately 250). For this subset, we extracted the abstracts from their respective PDFs. In order to train a classifier to identify connectivity information at the sentence level, it was necessary for us to manually markup a subset of our abstracts using the Knowtator annotation plugin for the Protege ontology management system

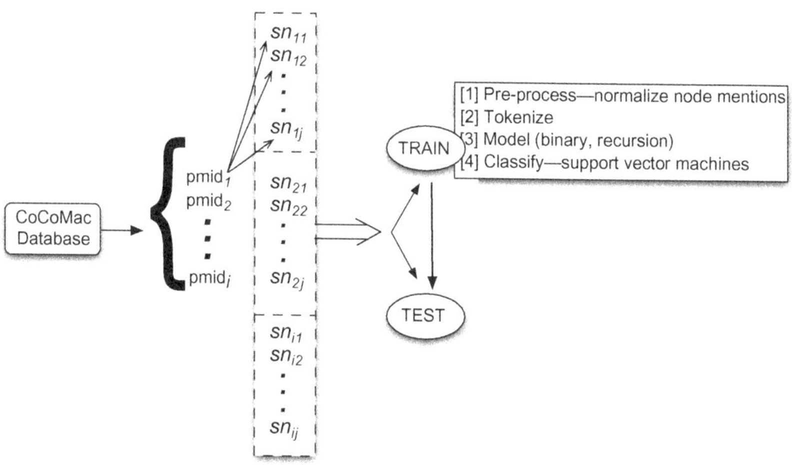

Figure 6.2 Workflow diagram of the classification system used in the present set of experiments. Full-text PDFs were obtained for the articles indexed in the CoCoMac Database, and each sentence within them was manually annotated as being positive or negative examples of a connection described in its associated CoCoMac entry. These sentences were then used to train a support vector machine-based classification system, using 5 × 2-way cross-validation.

(Ogren, 2006), identifying those sentences containing connectivity information, as well as any single- or multiword strings that refer to a particular neuroanatomical concept. For this proof of concept, we only annotated 60 articles in our data set; however, this resulted in a data set containing approximately 600 sentence/connectivity judgment pairs. We performed cross-validation on these data to develop a baseline SVM (Vapnik, 2000)-based classifier against which we compared the results of various feature selection and resampling experiments. For thoroughness, we compared the performance of our SVM-based systems to that of a non-SVM classifier, *k*IGNN, a mutual information-based *k*-nearest neighbor classifier that has been shown to be effective in identifying documents containing PPI-related information (Ambert & Cohen, 2011).

The performance of our baseline system, according to the area under the receiver operating characteristic (AUC), is depicted in Fig. 6.3. For the AUC, random classification would equate to a value of 0.5. Although our baseline system performs better than random (0.63 ± 0.05), an examination of the ratio of positive classes in light of previous research (Cohen, 2006) led us to hypothesize that the overabundance of negative class-sentences was leading to poor performance. To overcome this, we used a previously described resampling method (Cohen, 2006), in which we sampled (with replacement) from our existing data set to create a new one, but increased the probability that a given sample would be from the positive class. Performance of this approach is depicted in Fig. 6.3 for a range of probabilities for obtaining a positive class sample (1–5: 1x through 5x as likely). Importantly,

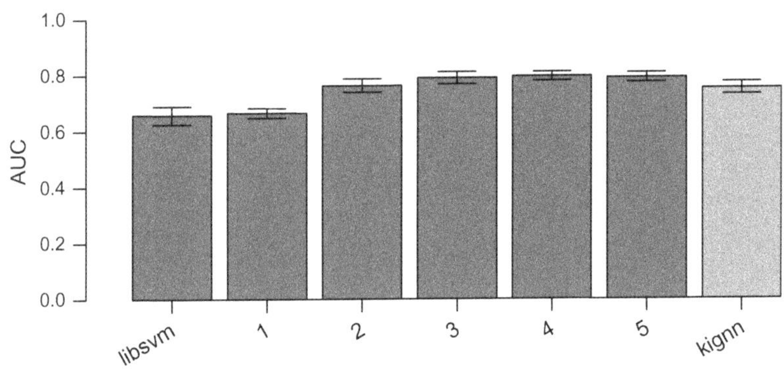

Figure 6.3 AUC (with 95% confidence intervals) comparisons of our baseline (libsvm) and various number of costs for misclassifying a positive sentence (1–5), with a previously successful relationship extraction system (kignn). (For color version of this figure, the reader is referred to the online version of this chapter.)

since this is a resampling method, even though the $1x$ probability level is equivalent to our baseline system, this method results in a data set five times as large as that of our baseline system. This is reflected in the fact that the AUC of the baseline and $1x$ system are roughly the same, but the $1x$ confidence intervals are much tighter.

We were interested in determining feature selection and feature generation methods that would lead to improved performance. Here, we examined the effects of neuroanatomical term normalization and neuroanatomical term-based distance feature generation on performance. Using the neuroanatomy markups obtained during our Knowtator annotation procedure, we replaced all recognized neuroanatomical features with a single common feature. To examine the effects of doing this on performance, we plotted the information gain associated with each feature for our normalized and non-normalized data sets (Fig. 6.4, normalized: blue; non-normalized: black). As this figure makes clear, when all neuroanatomical terms are replaced with a common feature, the peak of the information gain is sharper and shifted to the left. This implies that many of the predictive features in the non-normalized collection were neuroanatomical terms, and that performance would be improved by grouping all these into a single feature. In terms of qualitative implications, this would mean

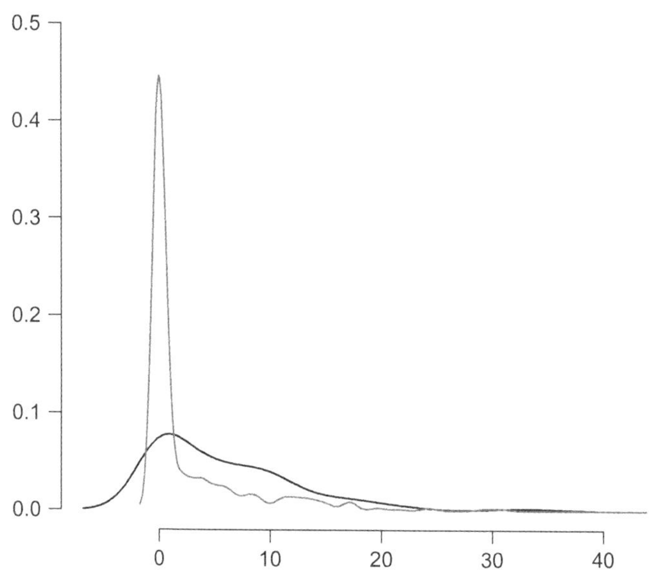

Figure 6.4 Distribution of average distance between neuroanatomical terms in the positive (black) and negative (red) classes. (For interpretation of the references to color in this figure legend, the reader is referred to the online version of this chapter.)

that one of the best ways our classification system was able to distinguish between sentences that were positive or negative for containing connectivity information was whether they contained neuroanatomical terms. Figure 6.4 depicts the distribution of the average distance between neuroanatomical terms within each sentence for the positive (black) and negative (red) classes. The results depicted in Fig. 6.5 fit well with those depicted here—the peak of the distribution for the negative class is sharply centered around 0 (meaning that one or fewer neuroanatomical terms were contained in the sentence). The positive class is also centered around 0, but it drops less gradually toward positive values. Based on these results, we hypothesized that normalizing our data set for neuroanatomical terms, as well as including a feature describing the average distance between neuroanatomical terms in a given sentence, would improve performance of our classifier. This combination of features led to substantial improvement in our cross-validation studies (AUC: 0.81).

This proof-of-concept text classification experiment demonstrates the feasibility of developing a sentence-level neuroanatomical relationship classifier using a small number of annotated articles. We were able to achieve a level of performance that could be useful for performing actual classification tasks (i.e., AUC $\geq$ 0.80) by using a SVM classifier and cost-based resampling methods. In practice, neuroscientists could use a system such as this to extract a literature-based *connectome* for a particular model organism. In particular, this tool could be integrated with a system recently developed by French and colleagues (French, 2012; French, Pavlidis, & Sporns, 2011) to identify specific brain regions and pull down their gene expression-related information from the Allen Brain Atlas (Lein et al., 2006). Integrating all this information could be used to create an integrated visual map of brain connections and their gene expression data that could be used, for example, to model spatial correlation of gene expressions in the brain.

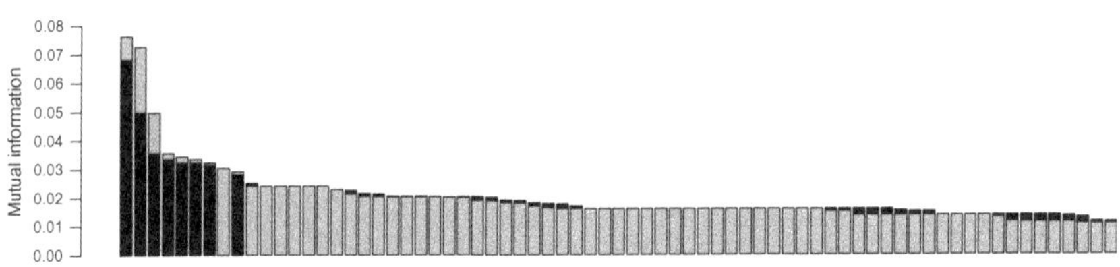

Figure 6.5 Feature information gain with (blue) and without (black) neuroanatomical term normalization. (For interpretation of the references to color in this figure legend, the reader is referred to the online version of this chapter.)

10. KNOWLEDGE MINING

One alternative to using machine learning for assisting manual database curation is that of automated mining from document databases. Because the financial and time costs associated with developing a large curated document collection are often prohibitive, researchers will sometimes perform automated association mining, in which textual features are extracted from a large collection of input documents and used either to further one's understanding of the relationships between the documents themselves or to develop hypotheses that can be investigated on their own. Voytek and Voytek (2012), for example, used co-occurrences of brain region mentions, cognitive functions, and brain-related diseases to demonstrate that known relationships can be extracted in an automated and scalable way by using clustering algorithms. Importantly, they were able to extend this approach to semi-automatically generate hypotheses regarding "holes" in the literature associations between brain structure and function, or function and disease which are likely to exist, but lack support in the literature. For example, they discovered that the structure *striatum* and the term *migraine* were strongly related to the term *serotonin* (they co-occurred in nearly 3000 publications for each relationship), yet the *striatum* and *migraine* had only 16 shared publications themselves, indicating that this association may exist but be understudied.

French and Pavlidis (2012) used knowledge mining to automatically map neuroanatomical identifiers found in a large volume of journal abstracts from the *Journal of Comparative Neurology* (*JCN*) to connect over 100,000 brain region mentions to 8225 normalized brain region concepts in a database. In this work, they used an annotated collection of abstracts from *JCN* and other neuroscience journals (French, Lane, Xu, & Pavlidis, 2009), expanding all abbreviations in the text, and manually identified the brain region mentions they contained. They also put together a dictionary of 7145 brain regions having formal unique identifiers from the NeuroNames vocabulary (Bowden et al., 2007), NIFSTD/BIRNLex (Bug et al., 2008), Brede Database (Nielsen, Hansen, & Balslev, 2004), Brain Architecture Management System (Bota & Swanson, 2008), and Allen Mouse Brain Reference Atlas (Dong, 2008). In total, they used five different techniques to link the free-text neuroanatomical mentions to the compiled set of terms: exact string matching, bag of words, stemming, bag of stems (similar to gap-edit global string matching; Srinivas, Cristianini, Jones, & Gorin, 2005), and the Lexical OWL Ontology Matcher, which allows for the specification of specific types of entities (Ghazvinian, Noy, & Musen, 2009). Scientists interested in using these resources could incorporate their annotated data (freely available at http://www.chibi.ubc.ca/WhiteText) into a classification system like the ones described in the previous section.

11. GRAND CHALLENGES AND FUTURE DIRECTIONS IN TEXT-MINING AND NEUROSCIENCE

As noted above, there are several current challenges in the field, at present, including developing new and improved data curation and data sharing methodologies. One area that has been getting some recent discussion is that of meaningful use of neuroscience metadata. In a recent editorial (Ascoli, 2012), Giorigio Ascoli emphasized the importance of tagging neuroscience publications with accurate metadata, for example, specific key words that will allow search engines to identify publications having data and results of interest to a reader. The advantage of this approach, over that of simple key word tagging, is that the metadata could be *a priori* mapped to ontology, allowing for more general queries (e.g., "Give me all documents that use a behavioral assay."). Going forward, one challenge will be to determine other metadata dimensions that further facilitate document retrieval, such as animal species, experimental methods used, or analytical techniques employed.

Although many resources, such as the NIF, are already available and actively used by research neuroscientists to query across data sources, there remains much work to be done. At present, a major emphasis in the field is developing tools that are easy to use and will foster data sharing and collaboration (Kennedy, 2012). One approach is to use social networking to connect authors who have complementary research topics (Bahr & Cohen, 2008) or by identifying scientists who share similar methods, but different research interests (Haendel, Vasilevsky, & Wirz, 2012). Future work will also include the development of tools that can make the development of curated data collections much more efficient. In our own lab, we are in the process of building an active learning recommender system that can be used to identify publications that contain information relevant to online collaboratively developed resources, such as the neuron registry or hippocampome (Ascoli, 2010; Hamilton, Shepherd, Martone, & Ascoli, 2012).

Even though its time in neuroscience is still young, text-mining has already had a substantial impact on the landscape of neuroscience research, and its importance will only continue to grow as the body of published literature increases. As organizations like the Allen Institute for Brain Science,

National Institutes of Health, and the International Neuroinformatics Coordinating Facility (INCF; http://www.incf.org/) continue to emphasize the importance of neuroscientific data integration, neuroinformaticians will increasingly rely on and extend the methodologies that have been described here, providing indirect benefit to behavioral researchers through the development of useful research utilities. More directly, behavioral scientists could adopt some of the procedures that have been described in this chapter to create their own repositories of literature relevant to their line of work or to mine such databases for gaps in our behavioral neuroscience knowledge.

REFERENCES

Ambert, K. H., & Cohen, A. M. (2009). A system for classifying disease comorbidity status from medical discharge summaries using automated hotspot and negated concept detection. *Journal of the American Medical Informatics Association*, *16*(4), 590 ISSN 1527-974X.

Ambert, K. H., & Cohen, A. M. (2011). k-Information gain scaled nearest neighbors: A novel approach to classifying protein-protein interactions in free-text. *IEEE Transaction on Computational Biology and Bioinformatics*, *9*(1), 305–310.

Ascoli, G. A. (2010). The coming of age of the hippocampome. *Neuroinformatics*, *8*(1), 1–3.

Ascoli, G. A. (2012). Twenty questions for neuroscience metadata. *Neuroinformatics*, *10*, 115–117.

Bahr, N. J., & Cohen, A. M. (2008). Discovering synergistic qualities of published authors to enhance translational research. In *AMIA Annual Symposium Proceedings 2008*. (p. 31). Washington D.C: American Medical Informatics Association.

Bandrowski, A. E. (2011). Biological resource catalog: NIF and NeuroLex. Available from Nature Precedings, http://dx.doi.org/10.1038/npre.2011.6238.1.

Baughman, R. W., Farkas, R., Guzman, M., & Huerta, M. F. (2006). The National Institutes of Health blueprint for neuroscience research. *The Journal of Neuroscience*, *26*(41), 10329–10331.

Bohland, J. W., Wu, C., Barbas, H., Bokil, H., Bota, M., Breiter, H. C., et al. (2009). A proposal for a coordinated effort for the determination of brainwide neuroanatomical connectivity in model organisms at a mesoscopic scale. *PLoS Computational Biology*, *5*, e1000334 Arxiv preprint arXiv:0901.4598.

Bota, M., & Swanson, L. W. (2007). The neuron classification problem. *Brain Research Reviews*, *56*(1), 79–88.

Bota, M., & Swanson, L. W. (2008). BAMS neuroanatomical ontology: Design and implementation. *Frontiers in Neuroinformatics*, *2*, 2.

Bowden, D. M., & Dubach, M. F. (2003). Neuronames 2002. *Neuroinformatics*, *1*(1), 43–59.

Bowden, D. M., Dubach, M., & Park, J. (2007). Creating neuroscience ontologies. *Methods in Molecular Biology*, *401*, 67.

Bowden, D. M., & Martin, R. F. (1995). NeuroNames brain hierarchy. *NeuroImage*, *2*(1), 63–83 ISSN 1053-8119.

Bug, W. J., Ascoli, G. A., Grethe, J. S., Gupta, A., Fennema-Notestine, C., Laird, A. R., et al. (2008). The nifstd and birnlex vocabularies: Building comprehensive ontologies for neuroscience. *Neuroinformatics*, *6*(3), 175–194.

Bult, C. J., Eppig, J. T., Kadin, J. A., Richardson, J. E., Blake, J. A., & Mouse Genome Database Group, (2008). The Mouse Genome Database (MGD): Mouse biology and model systems. *Nucleic Acids Research*, *36*(Suppl. 1), D724–D728.

Burns, G., Feng, D., & Hovy, E. (2008). Intelligent approaches to mining the primary research literature: Techniques, systems, and examples. In A. Kelemen, A. Abraham & Y. Liang (Eds.), *Computational intelligence in medical informatics*, Heidelberg: Springer Berlin 17–50.

Burns, G. A. P. C., Krallinger, M., Cohen, K., Wu, C. & Hirschman, L. (2009). Studying biocuration workflows. 3rd International biocuration conference, April 16, 2009.

Clarke, E., & Jacyna, L. S. (1987). *Nineteenth-century origins of neuroscientific concepts.* Berkley: University of California Press.

Cohen, A. M. (2006). An effective general purpose approach for automated biomedical document classification. In *AMIA annual symposium proceedings 2006.* (p. 161). Washington D.C: American Medical Informatics Association.

Cohen, A. M. (2008). Five-way smoking status classification using text hot-spot identification and error-correcting output codes. *Journal of the American Medical Informatics Association, 15*(1), 32–35.

Cohen, A. M., Adams, C. E., Davis, J. M., Yu, C., Yu, P. S., Meng, W., et al. (2010). Evidence-based medicine, the essential role of systematic reviews, and the need for automated text mining tools. In: *Proceedings of the 1st ACM international health informatics symposium* (pp. 376–380), New York City, NY: ACM.

Cohen, A. M., Ambert, K., & McDonagh, M. (2009). Cross-topic learning for work prioritization in systematic review creation and update. *Journal of the American Medical Informatics Association, 16*(5), 690–704.

Cohen, A. M., Ambert, K., Yang, J., Felder, R., Sproat, R., Roark, B., et al. (2010). OHSU/Portland VAMC team participation in the 2010 i2b2/VA challenge tasks. In: *Proceedings of the 2010 i2b2/VA workshop on challenges in natural language processing for clinical data, Boston, MA: i2b2.*

Cohen, A. M., & Hersh, W. R. (2005). A survey of current work in biomedical text mining. *Briefings in Bioinformatics, 6*(1), 57.

Cunningham, C. L., Gremel, C. M., & Groblewski, P. A. (2006). Drug-induced conditioned place preference and aversion in mice. *Nature Protocols, l*(4), 1662–1670.

Dong, H. W. (2008). *The Allen reference atlas: A digital color brain atlas of the C57Bl/6J male mouse.* San Francisco, CA: John Wiley & Sons.

French, L. H. (2012). Bioinformatics for neuroanatomical connectivity. http://hdl.handle.net/2429/40369.

French, L., Lane, S., Xu, L., & Pavlidis, P. (2009). Automated recognition of brain region mentions in neuroscience literature. *Frontiers in Neuroinformatics, 3*, 29.

French, L., & Pavlidis, P. (2007). Informatics in neuroscience. *Briefings in Bioinformatics, 8*, 446–456.

French, L., & Pavlidis, P. (2012). Using text mining to link journal articles to neuroanatomical databases. *The Journal of Comparative Neurology, 520*, 1772–1783.

French, L., Pavlidis, P., & Sporns, O. (2011). Relationships between gene expression and brain wiring in the adult rodent brain. *PLoS Computational Biology*, 7(1), 795–799 ISSN 1553-734X.

Gardner, D., Akil, H., Ascoli, G. A., Bowden, D. M., Bug, W., Donohue, D. E., et al. (2008). The neuroscience information framework: A data and knowledge environment for neuroscience. *Neuroinformatics, 6*(3), 149–160.

Ghazvinian, A., Noy, N. F., & Musen, M. A. (2009). Creating mappings for onto logies in biomedicine: Simple methods work. In *AMIA annual symposium proceedings 2009*, (p. 198). Washington D.C: American Medical Informatics Association.

Gupta, A., Bug, W., Marenco, L., Qian, X., Condit, C., Rangarajan, A., et al. (2008). Federated access to heterogeneous information resources in the neuroscience information framework (NIF). *Neuroinformatics, 6*(3), 205–217.

Haendel, M. A., Vasilevsky, N. A., & Wirz, J. A. (2012). Dealing with data: A case study on information and data management literacy. *PLoS Biology*, *10*(5), el001339.

Hamilton, D. J., Shepherd, G. M., Martone, M. E., & Ascoli, G. A. (2012). An ontological approach to describing neurons and their relationships. *Frontiers in Neuroinformatics*, *6*, 15.

Helmer, K. G., Ambite, J. L., Ames, J., Ananthakrishnan, R., Burns, G., Chervenak, A. L., et al. (2011). Enabling collaborative research using the biomedical informatics research network (BIRN). *Journal of the American Medical Informatics Association*, *18*(4), 416–422.

Hersh, W. R. (2009). *Information retrieval: A health and biomedical perspective*. New York, NY: Springer Verlag.

Hill, R. J., & Sternberg, P. W. (1992). The gene lin-3 encodes an inductive signal for vulval development in *C. elegans*. *Nature*, *358*(6386), 470.

Hirschman, L., Burns, G. A. P. C., Krallinger, M., Arighi, C., Cohen, K. B., Valencia, A., et al. (2012). Text mining for the biocuration workflow. *Database*, *2012*, http://dx.doi.org/10.1093/database/bas020.

Imam, F. T., Larson, S. D., Grethe, J. S., Gupta, A., Bandrowski, A., & Martone, M. E. (2012). Nifstd and neurolex: A comprehensive neuroscience ontology development based on multiple biomedical ontologies and community involvement. *Frontiers in Genetics*, *3*, 111.

Jensen, L. J., Saric, J., & Bork, P. (2006). Literature mining for the biologist: From information retrieval to biological discovery. *Nature Reviews Genetics*, *7*(2), 119–129.

Joachims, T. (1998). Text categorization with support vector machines: Learning with many relevant features. In: *Machine learning: ECML-98* (pp. 137–142).

Kennedy, D. N. (2012). The benefits of preparing data for sharing even when you don't. *Neuroinformatics*, *10*, 223–224.

Larson, S., Iman, F., Bakker, R., Pham, L., & Martone, M. (2010). A multi-scale parts list for the brain: Community-based ontology curation for neuroinformatics with NeuroLex.org. *Neuroinformatics*, http://www.frontiersin.org/10.3389/conf.fnins.2010.13.00079/event_abstract.

Larson, S. D., & Martone, M. E. (2009). Ontologies for neuroscience: What are they and what are they good for? *Frontiers in Neuroscience*, *3*(1), 60.

Lein, E. S., Hawrylycz, M. J., Ao, N., Ayres, M., Bensinger, A., Bernard, A., et al. (2006). Genome-wide atlas of gene expression in the adult mouse brain. *Nature*, *445*(7124), 168–176.

Martin, R. F., Dubach, J., & Bowden, D. (1990). Neuronames: Human/macaque neuroanatomical nomenclature. In: *Proceedings, UHTH annual symposium on computer applications in medical care* (pp. 1018–1019).

Martone, M. E., Gupta, A., & Ellisman, M. H. (2004). E-neuroscience: Challenges and triumphs in integrating distributed data from molecules to brains. *Nature Neuroscience*, *7*(5), 467–472.

Maynard, S. M., Mungall, C. J., Lewis, S. E., & Martone, M. E. (2010). A knowledge based approach to matching human neurodegenerative disease and associated animal models. *Neuroscience*, 230.

McCallum, A., & Nigam, K. (1998). A comparison of event models for naive Bayes text classification. In: *AAAI-98 workshop on learning for text categorization, Vol. 752* (pp. 41–48).

Mitchell, S. H., & Rosenthal, A. J. (2003). Effects of multiple delayed rewards on delay discounting in an adjusting amount procedure. *Behavioural Processes*, *64*(3), 273–286.

Müller, H. M., Kenny, E. E., & Sternberg, P. W. (2004). Textpresso: An ontology-based information retrieval and extraction system for biological literature. *PLoS Biology*, *2*(11), e309.

Müller, H. M., Rangarajan, A., Teal, T. K., & Sternberg, P. W. (2008). Textpresso for neuroscience: Searching the full text of thousands of neuroscience research papers. *Neuroinformatics*, *6*(3), 195–204.

Nielsen, F. A., Hansen, L. K., & Balslev, D. (2004). Mining for associations between text and brain activation in a functional neuroimaging database. *Neuroinformatics*, *2*(4), 369–379.

Ogren, P. V. (2006). Knowtator: A protégé plug-in for annotated corpus construction. In: *Proceedings of the 2006 conference of the North American chapter of the association for computational linguistics on human language technology: companion volume: demonstrations* (pp. 273–275), Sydney, Australia: Association for Computational Linguistics.

Pokkunuri, S., Ramakrishnan, C., Riloff, E., Hovy, E., & Burns, G. A. P. C. (2011). The role of information extraction in the design of a document triage application for biocuration. In: *Proceedings of BioNLP 2011 workshop* (pp. 46–55), Sydney, Australia: Association for Computational Linguistics.

PubMed Help. July 26th, (2012). http://www.ncbi.nlm.nih.gov/books/NBK3827/.

Ramakrishnan, C., Patnia, A., Hovy, E., Burns, G. A. P. C., Ramirez-Gonzalez, R. H., Bonnal, R., et al. (2012). Layout-aware text extraction from full-text pdf of scientific articles. *Source Code for Biology and Medicine*, 7(1), 7.

Rodgers, R. J., & Dalvi, A. (1997). Anxiety, defence and the elevated plus-maze. *Neuroscience and Biobehavioral Reviews*, *21*(6), 801–810.

Shepherd, G. M., Mirsky, J. S., Healy, M. D., Singer, M. S., Skoufos, E., Hines, M. S., et al. (1998). The Human Brain Project: Neuroinformatics tools for integrating, searching and modeling multidisciplinary neuroscience data. *Trends in Neurosciences*, *21*(11), 460–468 ISSN 0166-2236.

Sporns, O., Tononi, G., & Edelman, G. M. (2000). Theoretical neuroanatomy: Relating anatomical and functional connectivity in graphs and cortical connection matrices. *Cerebral Cortex*, *10*(2), 127–141.

Srinivas, P. R., Wei, S. H., Cristianini, N., Jones, E. G., & Gorin, F. A. (2005). Comparison of vector space model methodologies to reconcile cross-species neuroanatomical concepts. *Neuroinformatics*, *3*(2), 115–131.

Vapnik, V. N. (2000). *The nature of statistical learning theory*. New York, NY: Springer.

Voytek, J. B., & Voytek, B. (2012). Automated cognome construction and semi-automated hypothesis generation. *Journal of Neuroscience Methods*, *208*, 92–100.

Yang, J. J., Cohen, A. M., & McDonagh, M. S. (2008). Syriac: The systematic review information automated collection system a data warehouse for facilitating automated biomedical text classification. In: *AMIA Annual Symposium Proceedings*. *2008*, (p. 825).

CHAPTER SEVEN

Applying *In Silico* Integrative Genomics to Genetic Studies of Human Disease

Scott F. Saccone[1]

Department of Psychiatry, Washington University, Saint Louis, Missouri, USA

[1]Corresponding author: e-mail address: ssaccone@wustl.edu

Contents

Abstract

As genome-wide association studies using common single nucleotide polymorphism microarrays transition to whole-genome sequencing and the study of rare variants, new approaches will be required to viably interpret the results given the surge in data. A common strategy is to focus on biological hypotheses derived from sources of functional evidence ranging from the nucleotide to the biochemical process level. The accelerated development of biotechnology has led to numerous sources of functional evidence in the form of public databases and tools. Here, we review current methods and tools for integrating genomic data, particularly from the public domain, into genetic studies of human disease.

1. INTRODUCTION

Technological breakthroughs during the first decade of the twenty-first century led to a wave of discoveries in the mapping of human disease genes (Hindorff et al., 2009; Lander, 2011). High-throughput genotyping

International Review of Neurobiology, Volume 103
ISSN 0074-7742
http://dx.doi.org/10.1016/B978-0-12-388408-4.00007-1

on single nucleotide polymorphism (SNP) microarrays has been used in thousands of genome-wide association studies (GWAS) to identify numerous, independently replicated genotype–phenotype correlations for complex traits (Hardy & Singleton, 2009; Hindorff et al., 2009; Manolio, 2010). The success of GWAS was, however, tempered by observations that the variants discovered, which are mostly common (minor allele frequency greater than 5%), provided an incomplete picture of the genetic mechanisms underlying the traits (Goldstein, 2009; Hirschhorn, 2009). To complete the picture, investigators are using next-generation sequencing to study rare variants (Bahcall, 2012; Cirulli & Goldstein, 2010), copy number variation (CNV) (Conrad et al., 2009), and other forms of structural variation (Baker, 2012b). The challenges facing whole-genome disease mapping studies are now substantially greater given the potential loss of statistical power at rare variants (Ladouceur, Dastani, Aulchenko, Greenwood, & Richards, 2012) and the sheer size and complexity of these new datasets (McPherson, 2009).

By *in silico* integrative genomics, we mean the process of combining experimental data from multiple sources, such as association studies and external genomic resources, in an effort to discover a convergence of evidence from different experimental domains (Hawkins, Hon, & Ren, 2010). Because millions of genetic variants are tested for correlation with a phenotype, integrative methods are often used to focus the study by incorporating additional evidence for biological function (Hirschhorn, 2009). There are a number of issues to consider when applying integrative genomics to a genetic or translational genomic study. One is the determination of the experimental source of the data, or data provenance, and the assessment of its quality (Saccone, Quan, & Jones, 2012). Resources for integrative genomics rarely provide tools for systematically determining data provenance and assessing quality control. We provide some examples of new methods and tools that address these issues.

Another problem is how to measure the convergence of evidence. A standard tool for integrative genomics is the graphical genome browser which is used to visually inspect genomic data (Hawkins et al., 2010). While this method is easy to use and is very effective for studying small genomic regions, applications to whole-genome disease mapping studies can be problematic. The genome browser offers no quantitative measure of convergence and no reproducible algorithm for arriving at a conclusion, confounding factors such as linkage disequilibrium (LD) are difficult to incorporate, and it is difficult to automate. Automation is a major issue because in a whole-genome disease mapping study, using either a SNP microarray or whole-genome sequencing, integrative genomics can be used to identify functionally relevant variants among the thousands of those with nominal statistical significance, a task for which visual inspection in a genome browser is not viable. While the genome browser is a powerful tool for focusing on relatively small genomic intervals, other methods are required for whole-genome applications. We will review some algorithms and statistical methods used to integrate genetic and genomic data and assess convergence of evidence. We also discuss some tools that implement these methods on a genome-wide scale.

The continued growth of biotechnology will undoubtedly lead to further identification of variants that influence human disease and has the potential to determine their precise functional mechanisms—from transcription to protein to biochemical pathway. This will require substantial integration of genetic association studies with diverse genomic resources. Here, we review the current methods and tools for integrative genomics, how to assess data provenance and quality control and how to interpret the results.

2. GENOMIC RESOURCES

A useful hierarchy introduced by L. Stein (2001) divides genomic experimental data into three levels: the nucleotide, protein, and process levels. Experiments at the nucleotide level concern the observation of DNA and RNA, the transcription of DNA into RNA, the translation of RNA into protein, DNA–protein binding, and the regulation of transcription, as well as epigenetic structures. Protein level resources concern gene protein products and how genetic variants affect their structure. Process level data refer to the study of pathways and biochemical processes involving gene protein products. Protein and process level data are most readily used to hypothesize connections between phenotypes and genomic targets. Addiction, for example, could be studied by looking at genes whose protein products are in drug-related metabolic pathways (Li, Mao, & Wei, 2008) and then testing variation in these genes for association with the phenotype (Hinrichs et al., 2011).

Genomic resources at the nucleotide level include variation databases such as the HapMap (Frazer et al., 2007) and 1000 Genomes (Altshuler et al., 2010) projects, and dbSNP (Saccone et al., 2011; Sherry et al., 2001). These resources provide information on allele frequency estimates in various populations, maps of linkage disequilibrium, maps of genetic variants to gene transcripts, and what effect, if any, the variant has on the amino acid coding sequence such as missense and nonsense mutations. LD estimates are important for association studies because SNPs in high LD have correlated genotypes and therefore correlated association statistics. This is a major problem for disease mapping because it creates ambiguity in determining the true causal variant (Saccone, Saccone, Goate, et al., 2008; Ward & Kellis, 2012). Another important application of 1000 Genomes and HapMap data is genetic imputation which allows association studies to predict genotypes at untyped markers (Altshuler et al., 2010; Marchini & Howie, 2010). The dbSNP (Sherry et al., 2001) and dbVar (Sayers et al., 2011) databases at the National Center for Biotechnology Information (NCBI), as well as the Database of Genomic Variants (Zhang, Feuk, Duggan, Khaja, & Scherer, 2006), are major repositories for structural variation in numerous organisms, including humans. dbSNP provides a wide range of computational data such as mappings to reference genomes and gene transcripts and basic functional information on how variants affect transcription. Additional query and documentation tools for dbSNP are provided by the dbSNP-Q resource (Saccone et al., 2011). Information on CNVs can be found in the SCAN database (Gamazon et al., 2009), dbVar (Sayers et al., 2011), and the genetic variation database (Zhang et al., 2006). Cross-species sequence comparison can be used to identify potentially functional evolutionary conserved regions (ECRs) which are useful for studying noncoding regions (Bejerano et al., 2004; Loots et al., 2000; McCauley et al., 2007). Resources for ECR data include ECRbase (Loots & Ovcharenko, 2007) and the UCSC Genome Browser (Dreszer et al., 2011). General resources offering a wide range of experimental data and analytic tools at the nucleotide level include NCBI (Sayers et al., 2011), the UCSC Genome Browser (Dreszer et al., 2011; Rosenbloom et al., 2011), and Ensembl (Flicek et al., 2011). Much of the data from these resources can be systematically retrieved using tools such as Galaxy (Blankenberg, Coraor, Von Kuster, Taylor, & Nekrutenko, 2011) and BioMart (Guberman et al., 2011).

When a genetic variant appears to correlate with disease, a key question is whether there is additional evidence that the variant affects transcription. This is particularly important when numerous such variants from whole-genome experiments must be prioritized for further study. Polyphen-2 (Adzhubei et al., 2010), SIFT (Kumar, Henikoff, & Ng, 2009), and SNPdbe (Schaefer, Meier, Rost, & Bromberg, 2012) are resources for data on the predicted effects of amino acid changes. If the variants are in noncoding regions then regulatory data (Barnes & Plumpton, 2007; Chakravarti & Kapoor, 2012; Stormo, 2011) such as transcription factor binding sites and promoter regions can be studied using resources such as TRANSFAC (Wingender, 2008), ECRbase (Loots & Ovcharenko, 2007), and the UCSC Genome Browser (Dreszer et al., 2011)—particularly the UCSC implementation of the ENCODE data (Birney et al., 2007; Raney et al., 2010). Effects on transcription can also be studied by analyzing the correlation of variation with gene expression levels; variants with evidence of correlation are known as expression quantitative trait loci (eQTL) (Cookson, Liang, Abecasis, Moffatt, & Lathrop, 2009; Degner et al., 2012; Montgomery & Dermitzakis, 2011). From an integrative genomics perspective, eQTLs are attractive candidates for study as they provide a direct connection between human genetic variation and gene expression. Resources for eQTL data include the SCAN database (Nicolae et al., 2010) and the Pritchard lab (Degner et al., 2012). The GTEx project (http://commonfund.nih.gov/GTEx/) aims to provide eQTL resources by collecting expression data in several human tissues from densely genotyped subjects. Degner et al. (2012) linked variation in expression to epigenomics (Rakyan, Down, Balding, & Beck, 2011) by using DNase I sequencing to show that eQTLs are associated with chromatin accessibility. Epigenomic resources include the Pritchard Lab (Degner et al., 2012), the Human Epigenome Atlas (Bernstein et al., 2010), and the Human Epigenome Browser (Zhou et al., 2011).

Protein level resources such as UniPROT (Magrane & Consortium, 2011) provide data on the mapping of gene transcripts to proteins and how genetic variants affect protein structure. The Kyoto Encyclopedia of Genes and Genomes (Kanehisa, Goto, Sato, Furumichi, & Tanabe, 2011) provides a variety of protein and process level data including hierarchical classifications of genes and proteins and data on biochemical pathways. The Gene Ontology (GO) project (The Gene Ontology Consortium, 2011) provides highly structured data and tools that elucidate relationships between gene protein products and their biochemical functions.

Human disease resources include the database of genotypes and phenotypes (dbGaP) (Mailman et al., 2007), which provides GWAS genotype and phenotype data to qualified investigators, and the public NHGRI GWAS catalog (Hindorff et al., 2009), which provides selected GWAS results. Animal models including the Collaborative Cross (Churchill et al., 2004) can be used to study genes related to human disease, such as by studying patterns of gene expression in phenotyped mouse lines (Aylor et al., 2011); related data and tools can be found in the GeneNetwork (Wu, Huang, Juan, & Chen, 2004) and Mouse Genome Informatics resources (Blake, Bult, Kadin, Richardson, & Eppig, 2010; Finger et al., 2010). Animal models were one approach used in the NeuroSNP project (Saccone, Bierut, et al., 2009) to develop a database of genes and variants relevant to addiction-related phenotypes. Information on available knockout lines is available from the knockout mouse project (Austin et al., 2004). The NIMH Center for Collaborative Genomic Studies of Mental Disorders (http://nimhgenetics.org) provides genetic and deep phenotype data to qualified investigators of psychiatric disease and in some cases supplements the phenotypic data provided by dbGaP. Similarly, the NIDA Center for Genetic Studies (http://nidagenetics.org) provides data on addiction-related phenotypes. Biomaterials for subjects in the NIDA and NIMH repositories are provided to qualified investigators by the Rutgers University Cell and DNA repository (http://www.rucdr.org/).

3. METHODS OF INTEGRATIVE GENOMICS

3.1. Analytical frameworks

One of the early statistical approaches to integrative genomics introduced by Roeder, Devlin, and Wasserman (2007) used a weighting scheme that incorporated prior information in the form of external genomic data, such as gene expression in the brain for brain-disorder phenotypes. The weighting scheme would allow certain variants, such as those in expressed genes, to be weighted more heavily when assessing evidence of association from a GWAS. In terms of statistical power, the approach was shown to be robust to prior information that does not correlate with causal variants. While the study of genes expressed in the brain is a reasonable hypothesis to test, these genes do not necessarily contain causal variants. The weighting scheme was designed to minimize the loss of power resulting from uninformative prior information. Bayesian approaches to this problem include specifications of prior probabilities (Curtis, Vine, & Knight, 2007; Knight, Barnes, Breen, & Weale, 2011) and applications of hierarchical regression modeling (Chen & Witte, 2007; Lewinger, Conti, Baurley, Triche, & Thomas, 2007). While the evidence these integrative methods predict causal variants for human disease in general is still inconclusive, integrative genomics is often used to test biological hypotheses when limited resources require investigators to prioritize variants for further study (Baker, 2012a). Saccone, Saccone, Swan, et al. (2008), for example, developed the genomic information network (GIN) model for prioritizing GWAS results (Fig. 7.1). The GIN model is based on the weighting technique of Roeder et al. (2007) and is designed to viably integrate whole-genome data with maximum transparency and ease of interpretation. The GIN method has been implemented in the SPOT Web application (Saccone, Bolze, et al., 2010; Saccone, Culverhouse, et al., 2010) which is described in Section 3.2.

Another integrative approach for association studies is to look at the distribution of association statistics in certain classes of variants, such as those in certain biochemical pathways; this is sometimes referred to as gene set enrichment analysis (GSEA) (Wang, Li, & Hakonarson, 2010). Enrichment is useful not only for the identification of potential causal variants but also for the identification of biologically relevant pathways for disease (Hirschhorn, 2009). Holmans and colleagues introduced the ALIGATOR method for detecting enrichment of GWAS association signals in pathways from the GO database (Holmans et al., 2009). The ALIGATOR method corrects for LD, which causes ambiguity in the true causal variant; while an associated variant may be in a pathway, this may be due to LD with the true causal variant which may be tens of thousands of base pairs away and may reside in a different gene or not in a gene at all. Another factor that must be considered when evaluating the statistical significance of the findings is the size of the pathways; larger genes and larger numbers of genes will tend to contain a greater number of significant association statistics just due to chance. The ALIGATOR method corrects for this. Holmans and colleagues applied the ALIGATOR enrichment method to bipolar disorder. Other studies have reported pathways via GSEA analysis for bipolar (Smith et al., 2011), schizophrenia (O'Dushlaine et al., 2011; Richards et al., 2011), and autism spectrum disorder (ASD) (Voineagu et al., 2011; Wang, Zhang, et al., 2009). Enrichment analysis can also be applied to other forms of genomic data, such as eQTLs. Richards and colleagues found that Schizophrenia GWAS results are enriched for eQTLs and Nicolae and colleagues (Robinson et al., 2011) found eQTL enrichment using the NHGRI GWAS database (Hindorff et al., 2009).

3.2. Software

The Web-based graphical genome browser is arguably the most common integrative genomics tool (Hawkins et al., 2010). Figure 7.2 is a screenshot from the UCSC Genome Browser (Dreszer et al., 2011) showing a region on chromosome 15 associated with nicotine dependence (see Section 4).

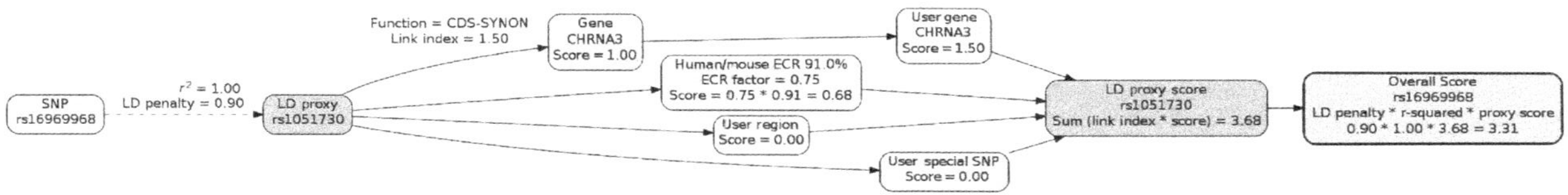

Figure 7.1 A genomic information network (GIN) from the SPOT Web application (*Saccone, Bolze, et al., 2010, with permission from Oxford University Press)* using the example data provided on the SPOT main page. Different sources of genomic data relating to a given SNP, *rs16969968*, are combined to form an overall measure of convergence of evidence or score. The score can be used to prioritize GWAS results for further study. Sources of evidence include SNP/transcript functional properties, predicted effects of missense mutations, evolutionary conservation, and user-defined candidate genes. In SPOT, the user can configure precisely how each type of data affects the score. The model takes into account LD estimated from a given HapMap population and will select the highest scoring LD correlated, or proxy, SNP. In this case, the PolyPhen prediction of "benign" for the missense SNP *rs16969968* in *CHRNA5* has led to the selection of the LD proxy coding SNP *rs1051730* in *CHRNA3* for determining the score of *rs16969968*. (For color version of this figure, the reader is referred to the online version of this chapter.)

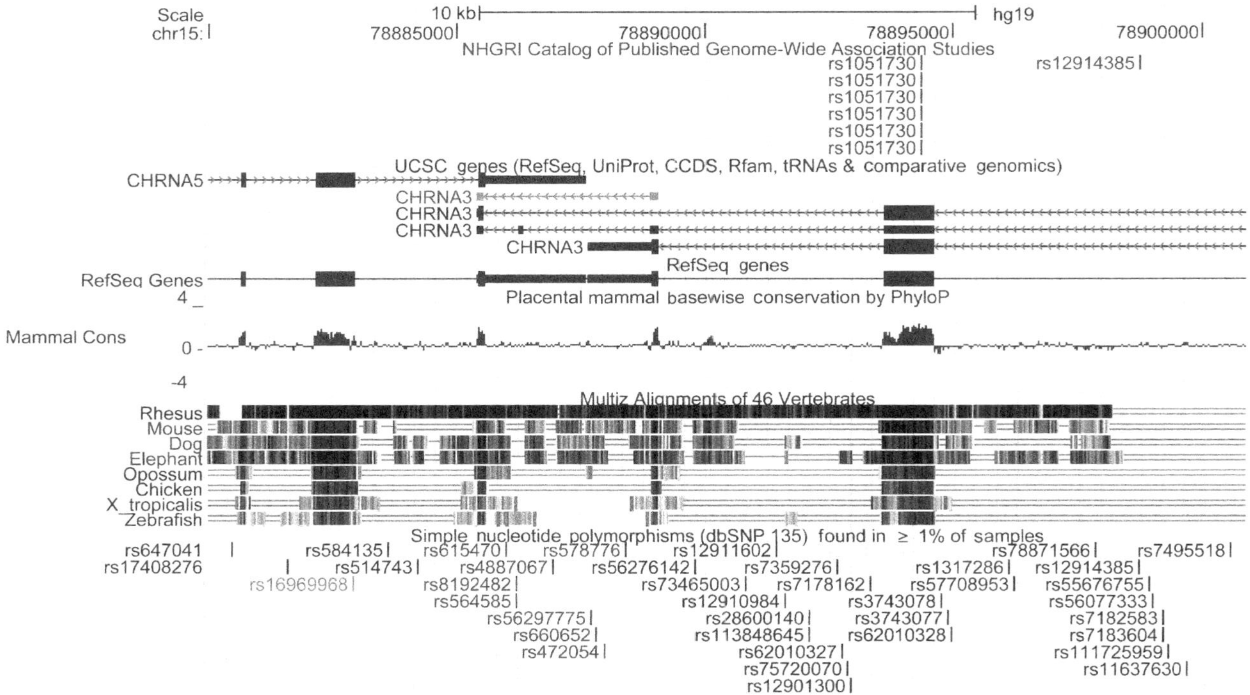

Figure 7.2 A view of a region on chromosome 15 in the UCSC genome browser showing GWAS results, gene transcripts, evolutionary conservation, and variants from dbSNP. The SNPs *rs16969968* and *rs1051730*, which are in complete LD ($r^2 = 1$ in the HapMap CEU sample), are associated with nicotine dependence and related phenotypes (see Section 4). (For color version of this figure, the reader is referred to the online version of this chapter.)

Other examples of genome browsers include the Generic Genome Browser (Donlin, 2007; Stein et al., 2002), Ensembl (Flicek et al., 2011), JBrowse (Westesson, Skinner, & Holmes, 2012), and the Human Epigenome Browser (Zhou et al., 2011). The UCSC and Ensembl resources in particular incorporate a vast array of cutting edge genomic databases. They allow investigators to download the underlying datasets and provide access to the data through a MySQL database server. In addition to Web-based genome browsers, a number of desktop applications are available such as the Integrated Genome Browser (Nicol, Helt, Blanchard, Raja, & Loraine, 2009), the Integrative Genomics Viewer (Robinson et al., 2011), and Savant (Fiume, Williams, Brook, & Brudno, 2010).

While genome browsers are easy to use, it is often difficult to extract the precise quantitative data underlying the graphical images and interpreting complex data such as LD patterns can be problematic. Furthermore, the method of visually assessing the convergence of genomic evidence is only viable for relatively small numbers of genetic variants and small genomic regions. A GWAS, for example, may lack the resources to pursue, through additional genotyping or functional studies, all variants with nominal non-genome-wide significance for association and may wish to prioritize thousands of variants for further study. The SPOT Web application (Saccone, Bolze, et al., 2010) accepts complete GWAS results and uses the GIN model (Saccone, Saccone, Swan, et al., 2008) (Section 3.1) to systematically rank the results by a quantitative measure of convergence of evidence from different genomic sources, including evidence for association provided by the investigator (Fig. 7.1). It accounts for ambiguity due to LD, variants being proximal to multiple genes and genes that have multiple reported transcripts. The SPOT implementation of the GIN model is not intended to be a predictive tool—the priorities for each type of genomic data can be set by the investigator and correspond to their specific genomic hypotheses. While genome browsers are very effective for focusing on relatively small genomic intervals and often incorporate more sources of genomic data, SPOT provides a more algorithmic and quantitative alternative to visual assessments of convergence that can be viably applied on a genome-wide scale.

For studies focusing on a particular variant, there are a number of tools that deal specifically with functional evidence such as the Variant Effect Predictor (VEP) (McLaren et al., 2010), PolyPhen2 (Adzhubei et al., 2010), SIFT (Kumar et al., 2009), and FastSNP (Yuan et al., 2006). VAAST (Yandell et al., 2011) combines a number of different strategies including known functional properties of variants as well as an analysis of Mendelian properties of alleles, including familial transmission data. When considering further experiments, such as functional studies focusing on a specific variant, it is important to determine if the variant of interest is in LD with other variants. HaploReg (Ward & Kellis, 2012) uses functional data to determine the most likely causal SNP among LD correlates.

Some tools deal mainly with genes and pathways rather than variation. DAVID (Huang da, Sherman, & Lempicki, 2009), for example, allows users to submit a list of genes, such as those containing GWAS hits. It will then perform an integrative enrichment analysis to determine if there are functional connections to biological processes and pathways among the set of genes. GeneWeaver (Baker, Jay, Bubier, Langston, & Chesler, 2012) and CANDID (Hutz, Kraja, McLeod, & Province, 2008) offer similar functionality. GRAIL (Raychaudhuri et al., 2009) applies literature mining data to a set of genomic regions and identifies functionally related genes and pathways.

3.3. Determining data provenance and assessing quality control

Integrative genomics is often used to make decisions of serious consequence. In genetic studies, it can guide the design of the study, such as selecting variants for follow-up experiments after an initial whole-genome association study. Follow-up experiments may involve sequencing or genotyping additional subjects, or costly functional studies using animal models (Bierut et al., 2008). In personalized medicine, a patient's genome may be cross-referenced with genomic databases to make diagnoses or enhance treatments (Calvo et al., 2012; Lyon, 2012). It is therefore important to determine the experimental source, or provenance, of the data and to assess quality control (Baggerly, 2010; Saccone et al., 2012). Underscoring this importance is a recent incident where erroneous genomic data were used as the basis of a cancer treatment study (Reich, 2011). This example is particularly poignant due to the numerous safeguards that were breached such as journal peer review, review by special committee, and published reports of irreproducibility (Coombes, Wang, & Baggerly, 2007). New safeguards are now being developed in response to this incident, including protocols for establishing data provenance (Duke Medicine Translational Medicine Quality Framework Committee, 2012).

Although provenance data for most genomic resources are made available to investigators, it is often difficult to locate, not well documented and typically obscured by an overwhelming assortment of visualization tools and external links. Furthermore, the task of conducting diagnostic quality control analyses, which is quite laborious due to the size of the datasets and the sheer number of issues that must be checked, is often left to the investigator. The BioQ Web application (Saccone et al., 2012) allows investigators to systematically assess data provenance for databases such as the 1000 Genomes project, HapMap, and dbSNP. Figure 7.3 is a screenshot from BioQ showing how frequency data from the 1000 Genomes project (Altshuler et al., 2010) can be traced back through a series of experiments and processes to the original subjects and biologics. The Biologic-Experiment-Result (BERT) data provenance model used in BioQ allows investigators to easily trace extensive information on experimental origins and measures of quality control. Additional models, such as FuGE (Jones & Lister, 2009), XCEDE (Gadde et al., 2011), and others (Zhao, Miles, Klyne, & Shotton, 2009), provide an increased level of experimental detail that may be more appropriate for specialized lab management and software development applications than direct use by general investigators.

4. APPLICATIONS

Whole-genome association studies of complex disease, either through a SNP microarray or whole-genome sequencing, are particularly challenging due to the high penalty for multiple testing (Chanock et al., 2007). This challenge can be mitigated, in some cases, by testing biological hypotheses based on the phenotype. One example is a study of nicotine dependence that used both GWAS (Bierut et al., 2007) and candidate gene (Saccone et al., 2007) designs. The candidate gene study focused on gene sets and biochemical pathways that were hypothesized to contain causal variants. A custom panel of SNPs was designed that ensured certain genes, such as nicotinic receptors, were more densely covered, and within these genes, exons and missense mutations were more highly prioritized. This *a priori* integrative genomics approach led to the discovery of a number of SNPs in the *CHRNA5–CHRNA3–CHRNB4* cluster of genes on chromosome 15, many of which were in strong LD (see Fig. 7.2). Of particular interest was a nonsynonymous SNP *rs16969968* in *CHRNA5*. Association at this SNP, along with its LD correlates, was later replicated in several other independent studies of nicotine dependence and related phenotypes such as cigarettes per day and heavy smoking (Amos, Spitz, & Cinciripini, 2010; Baker et al., 2009; Berrettini et al., 2008; Keskitalo et al., 2009; Saccone,

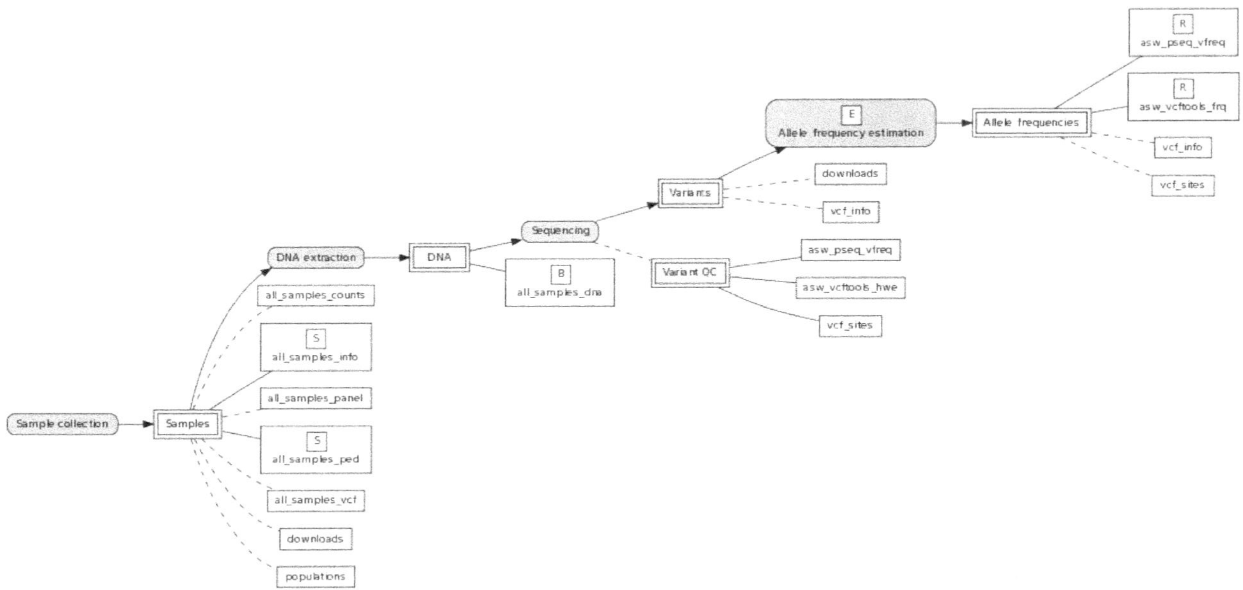

Figure 7.3 A screenshot from the BioQ Web application (*Saccone et al., 2012, with permission from Oxford University Press*) showing experimental process flow in the 1000 Genomes project. The Biologic-Experiment-Result (BERT) data provenance model is used to determine how allele frequency estimates (results—labeled "R") are traced back to the original subjects (labeled "S") and biologics (labeled "B"), such as DNA. The diagram is interactive in BioQ—selecting a node allows investigators to use query and documentation tools for detailed examination of the data. (For color version of this figure, the reader is referred to the online version of this chapter.)

Bierut, et al., 2009; Saccone, Culverhouse, et al., 2010; Saccone, Wang, et al., 2009; Sherva et al., 2008; Stevens et al., 2008; Thorgeirsson et al., 2008; Weiss et al., 2008), including a number of large meta-analytic studies (Furberg et al., 2010; Liu et al., 2010; Saccone, Culverhouse, et al., 2010; Thorgeirsson et al., 2010). These variants have also been reported to be associated with lung cancer (Amos et al., 2008; Hung et al., 2008; Liu et al., 2008; Thorgeirsson et al., 2008) and chronic obstructive pulmonary disease (Pillai et al., 2009). The missense SNP *rs16969968* in *CHRNA5* was shown *in vitro* to alter gene expression in mice (Bierut et al., 2008), and additional functional evidence was reported from a study of gene expression in the human brain (Wang, Cruchaga, et al., 2009). While this functional evidence suggests that the causal variants lie in the nicotinic receptor genes, there are LD-correlated SNPs in other genes, such as *IREB2* (DeMeo et al., 2009; Falvella et al., 2009) and *PSMA4* (Liu et al., 2009) that are also under investigation.

In a study of ASD, Voineagu et al. (2011) looked at patterns of gene expression in postmortem brain samples from 19 autism cases and 17 controls. Using coexpression network analysis, they found two network modules highly correlated with the phenotype. They then integrated the results with an ASD GWAS (Wang, Zhang, et al., 2009) and discovered significant enrichment for associations in one of these modules. While their sample of 36 subjects is somewhat small for the analysis of variation and eQTLs, the availability of human brain tissue allowed the investigators to discover novel biologically relevant targets that could be integrated into whole-genome association studies and applied to other neurodevelopmental diseases such as schizophrenia and attention deficit hyperactivity disorder. Another study (O'Dushlaine et al., 2011) found schizophrenia and bipolar GWAS results to be enriched in cell adhesion molecule pathways, which contain genes implicated in the same ASD GWAS (Wang, Zhang, et al., 2009) used by Voineagu and colleagues. A whole-genome ASD study of CNV (Pinto et al., 2010) found the results to be enriched in gene sets involved in GTPase/Ras signaling as well as microtubule cytoskeleton, glycosylation, and CNS development and adhesion (Wegiel et al., 2010). Overall, these integrative enrichment and pathway-based approaches suggest compelling biological hypotheses for the genetic study of neurodevelopmental psychiatric disorders.

Molecular diagnosis is another potential application of integrative genomics. A recent pilot study by Calvo et al. (2012) used integrative prioritization techniques to develop new methods of diagnosing human oxidative phosphorylation (OXPHOS) disease. Genetic analysis of this rare condition is complicated by numerous factors including both clinical and genetic heterogeneity, multiple inheritance models, pleiotropy, and genetic effects stemming from both nuclear and mitochondrial genes. These challenges are compounded by large numbers of rare variants in known gene targets. Focusing on mitochondrial targets, Calvo and colleagues developed a technique that prioritizes variants using various sources of functional evidence such as missense mutations predicted to be deleterious and evolutionary conservation. The study found that, of the 42 cases sequenced, 31% were due to novel, rare, recessive missense mutations and 24% were due to known mutations, while the remaining 45% could not be explained by molecular diagnosis. While the molecular approach is not yet the definitive diagnostic tool for OXPHOS, this pilot study underscores the potential for these integrative methods to be used as a diagnostic tool.

5. DISCUSSION

One issue for interpreting these methods is whether integrative genomics can be used to reduce the penalty for multiple testing when determining statistical significance by restricting to variants with certain properties such as those in candidate genes. A problem with this approach is that it is not difficult to contrive *post hoc* justifications for focusing on certain genes. In the study of addiction, for example, an abundance of pathways makes it relatively easy to find variants of nominal significance in genes from these pathways and so a reduced correction for multiple testing will lead to false positives. Caution must therefore be used in setting thresholds other than conventional genome-wide thresholds such as $p < 5 \times 10^{-7}$ (The Wellcome Trust Case Control Consortium, 2007), particularly if this is not clearly declared prior to analysis (Chanock et al., 2007). This threshold can of course be relaxed when it is being used to select variants for further study, such as sequencing additional samples to provide greater statistical power and increased significance of association findings.

A key problem for integrative genomics is to assess the extent to which external genomic data from public resources will increase the chances of identifying a true causal variant, that is, to what extent the process of integrative genomics is predictive. A fundamental issue is to identify the outcome being predicted. Human disease in general is an intractably broad outcome. This is particularly problematic for nucleotide level data where there is no clear connection to any general class of disease traits. Gene expression, when considered as a quantitative trait, is one alternative; this is the study of eQTLs (Section 2). The study of biochemical pathways for certain classes of complex disease such as addiction and neurodevelopmental diseases has provided some promising results (Section 4), although it is not clear the success of this approach generalizes to a broader class of conditions. LD is another major issue in assessing the performance of integrative genomics because it creates ambiguity in true causal variant. The variants discovered for nicotine dependence discussed in Section 4, for example, are in strong LD with variants extending across several genes, and there is currently no definitive functional evidence that identifies the true causal variants. This is a challenging issue that cannot be resolved by sequencing additional subjects in the same population because LD patterns will continue to cause ambiguity. While cross-population (Saccone, Saccone, Goate, et al., 2008) and other methods (Ward & Kellis, 2012) have had some success in narrowing the evidence among LD correlates, additional evidence is required to definitively resolve this issue. Ultimately, establishing the extent to which these methods predict causal variants for general human disease may require a set of confirmed, independently replicated, LD-disambiguated association results large enough for a viable statistical analysis, which is clearly a major challenge.

There are some methods involving mainly nucleotide level data that are based on direct connections to general human disease. One example is the PolyPhen (Adzhubei et al., 2010) method of predicting the impact of amino acid substitutions. Part of the predictive training for this model involves deleterious mutations with major detrimental effects such as complete loss of function (LoF) and death, and therefore, care must be taken in applications to common complex disease; the PolyPhen2 software package does provide options for dealing with this issue. Recent studies examining the genomes of seemingly healthy human subjects predict them to have more than 200 LoF variants (Altshuler et al., 2010; Ng et al., 2008), with some studies predicting as many as 800 (MacArthur et al., 2012; Pelak et al., 2010). Therefore, evidence appearing to suggest a substantial functional effect may have a less than expected phenotypic impact. Hindorff et al. (2009) noted that a substantial number of GWAS results lack clear evidence of a functional effect based on variant/transcript properties. Of the 531 genome-wide significant GWAS results they considered, 45% were intronic and 43%

were intergenic. This underscores the value of whole-genome approaches which may harbor the potential to discover new functional mechanisms.

While several genetic studies have successfully used integrative genomics to test biologically compelling hypotheses, the extent to which these approaches quantitatively predict causal variants remains unclear. Prioritizing experiments by testing biologically compelling hypotheses is nevertheless a reasonable approach (Baker, 2012a), particularly when resources are limited. While there are many resources for using integrative genomics, there are also many issues to consider. Future tools will most likely provide greater clarity in the source and quality of genomic data and an improved means of making connections to the phenotype.

ACKNOWLEDGMENT

This work was supported by a grant from the National Institute on Drug Abuse (K01DA024722).

REFERENCES

Adzhubei, I. A., Schmidt, S., Peshkin, L., Ramensky, V. E., Gerasimova, A., Bork, P., et al. (2010). A method and server for predicting damaging missense mutations. *Nature Methods*, *7*, 248–249.

Altshuler, D. M., Gibbs, R. A., Peltonen, L., Dermitzakis, E., Schaffner, S. F., Yu, F., et al. (2010). Integrating common and rare genetic variation in diverse human populations. *Nature*, *467*, 52–58.

Amos, C. I., Spitz, M. R., & Cinciripini, P. (2010). Chipping away at the genetics of smoking behavior. *Nature Genetics*, *42*, 366–368.

Amos, C. I., Wu, X., Broderick, P., Gorlov, I. P., Gu, J., Eisen, T., et al. (2008). Genome-wide association scan of tag SNPs identifies a susceptibility locus for lung cancer at 15q25.1. *Nature Genetics*, *40*, 616–622.

Austin, C. P., Battey, J. F., Bradley, A., Bucan, M., Capecchi, M., Collins, F. S., et al. (2004). The knockout mouse project. *Nature Genetics*, *36*, 921–924.

Aylor, D. L., Valdar, W., Foulds-Mathes, W., Buus, R. J., Verdugo, R. A., Baric, R. S., et al. (2011). Genetic analysis of complex traits in the emerging Collaborative Cross. *Genome Research*, *21*, 1213–1222.

Baggerly, K. (2010). Disclose all data in publications. *Nature*, *467*, 401.

Bahcall, O. (2012). Rare variant association. *Nature Genetics*, *44*, 241.

Baker, M. (2012a). Functional genomics: The changes that count. *Nature*, *482*(257), 259–262.

Baker, M. (2012b). Structural variation: The genome's hidden architecture. *Nature Methods*, *9*, 133–137.

Baker, E. J., Jay, J. J., Bubier, J. A., Langston, M. A., & Chesler, E. J. (2012). GeneWeaver: A web-based system for integrative functional genomics. *Nucleic Acids Research*, *40*, D1067–D1076.

Baker, T. B., Weiss, R. B., Bolt, D., von Niederhausern, A., Fiore, M. C., Dunn, D. M., et al. (2009). Human neuronal acetylcholine receptor A5-A3-B4 haplotypes are associated with multiple nicotine dependence phenotypes. *Nicotine & Tobacco Research*, *11*, 785–796.

Barnes, M. R., & Plumpton, M. (2007). Predictive functional analysis of polymorphisms: An overview. In M. R. Barnes (Ed.), *Bioinformatics for geneticists* (pp. 249–280). England: John Wiley & Sons, Ltd., West Sussex PO19 8SQ.

Bejerano, G., Pheasant, M., Makunin, I., Stephen, S., Kent, W. J., Mattick, J. S., et al. (2004). Ultraconserved elements in the human genome. *Science, 304,* 1321–1325.

Bernstein, B. E., Stamatoyannopoulos, J. A., Costello, J. F., Ren, B., Milosavljevic, A., Meissner, A., et al. (2010). The NIH roadmap epigenomics mapping consortium. *Nature Biotechnology, 28,* 1045–1048.

Berrettini, W., Yuan, X., Tozzi, F., Song, K., Francks, C., Chilcoat, H., et al. (2008). alpha-5/alpha-3 nicotinic receptor subunit alleles increase risk for heavy smoking. *Molecular Psychiatry, 13,* 368–373.

Bierut, L. J., Madden, P. A., Breslau, N., Johnson, E. O., Hatsukami, D., Pomerleau, O. F., et al. (2007). Novel genes identified in a high-density genome wide association study for nicotine dependence. *Human Molecular Genetics, 16,* 24–35.

Bierut, L. J., Stitzel, J. A., Wang, J. C., Hinrichs, A. L., Grucza, R. A., Xuei, X., et al. (2008). Variants in nicotinic receptors and risk for nicotine dependence. *The American Journal of Psychiatry, 165,* 1163–1171.

Birney, E., Stamatoyannopoulos, J. A., Dutta, A., Guigo, R., Gingeras, T. R., Margulies, E. H., et al. (2007). Identification and analysis of functional elements in 1% of the human genome by the ENCODE pilot project. *Nature, 447,* 799–816.

Blake, J. A., Bult, C. J., Kadin, J. A., Richardson, J. E., & Eppig, J. T. (2010). The mouse genome database (MGD): Premier model organism resource for mammalian genomics and genetics. *Nucleic Acids Research, 39,* D842–D848.

Blankenberg, D., Coraor, N., Von Kuster, G., Taylor, J., & Nekrutenko, A. (2011). Integrating diverse databases into an unified analysis framework: A Galaxy approach. *Database: The Journal of Biological Databases and Curation, 2011,* bar011.

Calvo, S. E., Compton, A. G., Hershman, S. G., Lim, S. C., Lieber, D. S., Tucker, E. J., et al. (2012). Molecular diagnosis of infantile mitochondrial disease with targeted next-generation sequencing. *Science Translational Medicine, 4,* 118ra110.

Chakravarti, A., & Kapoor, A. (2012). Genetics. Mendelian puzzles. *Science, 335,* 930–931.

Chanock, S. J., Manolio, T., Boehnke, M., Boerwinkle, E., Hunter, D. J., Thomas, G., et al. (2007). Replicating genotype-phenotype associations. *Nature, 447,* 655–660.

Chen, G. K., & Witte, J. S. (2007). Enriching the analysis of genomewide association studies with hierarchical modeling. *American Journal of Human Genetics, 81,* 397–404.

Churchill, G. A., Airey, D. C., Allayee, H., Angel, J. M., Attie, A. D., Beatty, J., et al. (2004). The Collaborative Cross, a community resource for the genetic analysis of complex traits. *Nature Genetics, 36,* 1133–1137.

Cirulli, E. T., & Goldstein, D. B. (2010). Uncovering the roles of rare variants in common disease through whole-genome sequencing. *Nature Reviews Genetics, 11,* 415–425.

Conrad, D. F., Pinto, D., Redon, R., Feuk, L., Gokcumen, O., Zhang, Y., et al. (2009). Origins and functional impact of copy number variation in the human genome. *Nature, 464,* 704–712.

Cookson, W., Liang, L., Abecasis, G., Moffatt, M., & Lathrop, M. (2009). Mapping complex disease traits with global gene expression. *Nature Reviews Genetics, 10,* 184–194.

Coombes, K. R., Wang, J., & Baggerly, K. A. (2007). Microarrays: Retracing steps. *Nature Medicine, 13,* 1276–1277 author reply 1277–1278.

Curtis, D., Vine, A. E., & Knight, J. (2007). A pragmatic suggestion for dealing with results for candidate genes obtained from genome wide association studies. *BMC Genetics, 8,* 20.

Degner, J. F., Pai, A. A., Pique-Regi, R., Veyrieras, J. B., Gaffney, D. J., Pickrell, J. K., et al. (2012). DNase I sensitivity QTLs are a major determinant of human expression variation. *Nature, 482,* 390–394.

DeMeo, D. L., Mariani, T., Bhattacharya, S., Srisuma, S., Lange, C., Litonjua, A., et al. (2009). Integration of genomic and genetic approaches implicates IREB2 as a COPD susceptibility gene. *American Journal of Human Genetics, 85*, 493–502.

Donlin, M. J. (2007). Using the Generic Genome Browser (GBrowse). *Current Protocols in Bioinformatics*, Chapter 9, Unit 9 9.

Dreszer, T. R., Karolchik, D., Zweig, A. S., Hinrichs, A. S., Raney, B. J., Kuhn, R. M., et al. (2011). The UCSC Genome Browser database: Extensions and updates 2011. *Nucleic Acids Research, 40*, D918–D923.

Duke Medicine Translational Medicine Quality Framework Committee, (2012). *A framework for the quality of translational medicine with a focus on human genomic studies*. http://medschool.duke.edu/files/Translational_Medicine_Quality_Framework_Principles_-_May_1%2C_2011%5B1%5D.pdf . Retrieved March 15, 2012.

Falvella, F. S., Galvan, A., Frullanti, E., Spinola, M., Calabro, E., Carbone, A., et al. (2009). Transcription deregulation at the 15q25 locus in association with lung adenocarcinoma risk. *Clinical Cancer Research, 15*, 1837–1842.

Finger, J. H., Smith, C. M., Hayamizu, T. F., McCright, I. J., Eppig, J. T., Kadin, J. A., et al. (2010). The mouse Gene Expression Database (GXD): 2011 update. *Nucleic Acids Research, 39*, D835–D841.

Fiume, M., Williams, V., Brook, A., & Brudno, M. (2010). Savant: Genome browser for high-throughput sequencing data. *Bioinformatics, 26*, 1938–1944.

Flicek, P., Amode, M. R., Barrell, D., Beal, K., Brent, S., Carvalho-Silva, D., et al. (2011). Ensembl 2012. *Nucleic Acids Research, 40*, D84–D90.

Frazer, K. A., Ballinger, D. G., Cox, D. R., Hinds, D. A., Stuve, L. L., Gibbs, R. A., et al. (2007). A second generation human haplotype map of over 3.1 million SNPs. *Nature, 449*, 851–861.

Furberg, H., Kim, Y., Dackor, J., Boerwinkle, E., Franceschini, N., Ardissino, D., et al. (2010). Genome-wide meta-analyses identify multiple loci associated with smoking behavior. *Nature Genetics, 42*, 441–447.

Gadde, S., Aucoin, N., Grethe, J. S., Keator, D. B., Marcus, D. S., & Pieper, S. (2011). XCEDE: An extensible schema for biomedical data. *Neuroinformatics, 10*, 19–32.

Gamazon, E. R., Zhang, W., Konkashbaev, A., Duan, S., Kistner, E. O., Nicolae, D. L., et al. (2009). SCAN: SNP and copy number annotation. *Bioinformatics, 26*, 259–262.

Goldstein, D. B. (2009). Common genetic variation and human traits. *The New England Journal of Medicine, 360*, 1696–1698.

Guberman, J. M., Ai, J., Arnaiz, O., Baran, J., Blake, A., Baldock, R., et al. (2011). BioMart Central Portal: An open database network for the biological community. *Database: The Journal of Biological Databases and Curation, 2011*, bar041.

Hardy, J., & Singleton, A. (2009). Genomewide association studies and human disease. *The New England Journal of Medicine, 360*, 1759–1768.

Hawkins, R. D., Hon, G. C., & Ren, B. (2010). Next-generation genomics: An integrative approach. *Nature Reviews Genetics, 11*, 476–486.

Hindorff, L. A., Sethupathy, P., Junkins, H. A., Ramos, E. M., Mehta, J. P., Collins, F. S., et al. (2009). Potential etiologic and functional implications of genome-wide association loci for human diseases and traits. *Proceedings of the National Academy of Sciences of the United States of America, 106*, 9362–9367.

Hinrichs, A. L., Murphy, S. E., Wang, J. C., Saccone, S., Saccone, N., Steinbach, J. H., et al. (2011). Common polymorphisms in FMO1 are associated with nicotine dependence. *Pharmacogenetics and Genomics, 21*, 397–402.

Hirschhorn, J. N. (2009). Genomewide association studies—Illuminating biologic pathways. *The New England Journal of Medicine, 360*, 1699–1701.

Holmans, P., Green, E. K., Pahwa, J. S., Ferreira, M. A., Purcell, S. M., Sklar, P., et al. (2009). Gene ontology analysis of GWA study data sets provides insights into the biology of bipolar disorder. *American Journal of Human Genetics, 85*, 13–24.

Huang da, W., Sherman, B. T., & Lempicki, R. A. (2009). Systematic and integrative analysis of large gene lists using DAVID bioinformatics resources. *Nature Protocols*, *4*, 44–57.

Hung, R. J., McKay, J. D., Gaborieau, V., Boffetta, P., Hashibe, M., Zaridze, D., et al. (2008). A susceptibility locus for lung cancer maps to nicotinic acetylcholine receptor subunit genes on 15q25. *Nature*, *452*, 633–637.

Hutz, J. E., Kraja, A. T., McLeod, H. L., & Province, M. A. (2008). CANDID: A flexible method for prioritizing candidate genes for complex human traits. *Genetic Epidemiology*, *32*, 779–790.

Jones, A. R., & Lister, A. L. (2009). Managing experimental data using FuGE. *Methods in Molecular Biology*, *604*, 333–343.

Kanehisa, M., Goto, S., Sato, Y., Furumichi, M., & Tanabe, M. (2011). KEGG for integration and interpretation of large-scale molecular data sets. *Nucleic Acids Research*, *40*, D109–D114.

Keskitalo, K., Broms, U., Heliövaara, M., Ripatti, S., Surakka, I., Perola, M., et al. (2009). Association of serum cotinine level with a cluster of three nicotinic acetylcholine receptor genes (CHRNA3/CHRNA5/CHRNB4) on chromosome 15. *Human Molecular Genetics*, *18*, 4007–4012.

Knight, J., Barnes, M. R., Breen, G., & Weale, M. E. (2011). Using functional annotation for the empirical determination of Bayes factors for genome-wide association study analysis. *PLoS One*, *6*, e14808.

Kumar, P., Henikoff, S., & Ng, P. C. (2009). Predicting the effects of coding non-synonymous variants on protein function using the SIFT algorithm. *Nature Protocols*, *4*, 1073–1081.

Ladouceur, M., Dastani, Z., Aulchenko, Y. S., Greenwood, C. M., & Richards, J. B. (2012). The empirical power of rare variant association methods: Results from sanger sequencing in 1,998 individuals. *PLoS Genetics*, *8*, e1002496.

Lander, E. S. (2011). Initial impact of the sequencing of the human genome. *Nature*, *470*, 187–197.

Lewinger, J. P., Conti, D. V., Baurley, J. W., Triche, T. J., & Thomas, D. C. (2007). Hi erarchical Bayes prioritization of marker associations from a genome-wide association scan for further investigation. *Genetic Epidemiology*, *31*, 871–882.

Li, C. Y., Mao, X., & Wei, L. (2008). Genes and (common) pathways underlying drug addiction. *PLoS Computational Biology*, *4*, e2.

Liu, Y., Liu, P., Wen, W., James, M. A., Wang, Y., Bailey-Wilson, J. E., et al. (2009). Haplotype and cell proliferation analyses of candidate lung cancer susceptibility genes on chromosome 15q24-25.1. *Cancer Research*, *69*, 7844–7850.

Liu, J. Z., Tozzi, F., Waterworth, D. M., Pillai, S. G., Muglia, P., Middleton, L., et al. (2010). Meta-analysis and imputation refines the association of 15q25 with smoking quantity. *Nature Genetics*, *42*, 436–440.

Liu, P., Vikis, H. G., Wang, D., Lu, Y., Wang, Y., Schwartz, A. G., et al. (2008). Familial aggregation of common sequence variants on 15q24-25.1 in lung cancer. *Journal of the National Cancer Institute*, *100*, 1326–1330.

Loots, G. G., Locksley, R. M., Blankespoor, C. M., Wang, Z. E., Miller, W., Rubin, E. M., et al. (2000). Identification of a coordinate regulator of interleukins 4, 13, and 5 by cross-species sequence comparisons. *Science*, *288*, 136–140.

Loots, G., & Ovcharenko, I. (2007). ECRbase: Database of evolutionary conserved regions, promoters, and transcription factor binding sites in vertebrate genomes. *Bioinformatics*, *23*, 122–124.

Lyon, G. J. (2012). Personalized medicine: Bring clinical standards to human-genetics research. *Nature*, *482*, 300–301.

MacArthur, D. G., Balasubramanian, S., Frankish, A., Huang, N., Morris, J., Walter, K., et al. (2012). A systematic survey of loss-of-function variants in human protein-coding genes. *Science*, *335*, 823–828.

Magrane, M., & Consortium, The UniProt (2011). UniProt Knowledgebase: A hub of integrated protein data. *Database: The Journal of Biological Databases and Curation*, *2011*, bar009.

Mailman, M. D., Feolo, M., Jin, Y., Kimura, M., Tryka, K., Bagoutdinov, R., et al. (2007). The NCBI dbGaP database of genotypes and phenotypes. *Nature Genetics*, *39*, 1181–1186.

Manolio, T. A. (2010). Genomewide association studies and assessment of the risk of disease. *The New England Journal of Medicine*, *363*, 166–176.

Marchini, J., & Howie, B. (2010). Genotype imputation for genome-wide association studies. *Nature Reviews Genetics*, *11*, 499–511.

McCauley, J. L., Kenealy, S. J., Margulies, E. H., Schnetz-Boutaud, N., Gregory, S. G., Hauser, S. L., et al. (2007). SNPs in Multi-Species Conserved Sequences (MCS) as useful markers in association studies: A practical approach. *BMC Genomics*, *8*, 266.

McLaren, W., Pritchard, B., Rios, D., Chen, Y., Flicek, P., & Cunningham, F. (2010). Deriving the consequences of genomic variants with the Ensembl API and SNP Effect Predictor. *Bioinformatics*, *26*, 2069–2070.

McPherson, J. D. (2009). Next-generation gap. *Nature Methods*, *6*, S2–S5.

Montgomery, S. B., & Dermitzakis, E. T. (2011). From expression QTLs to personalized transcriptomics. *Nature Reviews Genetics*, *12*, 277–282.

Ng, P. C., Levy, S., Huang, J., Stockwell, T. B., Walenz, B. P., Li, K., et al. (2008). Genetic variation in an individual human exome. *PLoS Genetics*, *4*, e1000160.

Nicol, J. W., Helt, G. A., Blanchard, S. G., Raja, A., & Loraine, A. E. (2009). The Integrated Genome Browser: Free software for distribution and exploration of genome-scale data sets. *Bioinformatics*, *25*, 2730–2731.

Nicolae, D. L., Gamazon, E., Zhang, W., Duan, S., Dolan, M. E., & Cox, N. J. (2010). Trait-associated SNPs are more likely to be eQTLs: Annotation to enhance discovery from GWAS. *PLoS Genetics*, *6*, e1000888.

O'Dushlaine, C., Kenny, E., Heron, E., Donohoe, G., Gill, M., Morris, D., et al. (2011). Molecular pathways involved in neuronal cell adhesion and membrane scaffolding contribute to schizophrenia and bipolar disorder susceptibility. *Molecular Psychiatry*, *16*, 286–292.

Pelak, K., Shianna, K. V., Ge, D., Maia, J. M., Zhu, M., Smith, J. P., et al. (2010). The characterization of twenty sequenced human genomes. *PLoS Genetics*, *6*, e1001111.

Pillai, S. G., Ge, D., Zhu, G., Kong, X., Shianna, K. V., Need, A. C., et al. (2009). A genome-wide association study in chronic obstructive pulmonary disease (COPD): Identification of two major susceptibility loci. *PLoS Genetics*, *5*, e1000421.

Pinto, D., Pagnamenta, A. T., Klei, L., Anney, R., Merico, D., Regan, R., et al. (2010). Functional impact of global rare copy number variation in autism spectrum disorders. *Nature*, *466*, 368–372.

Rakyan, V. K., Down, T. A., Balding, D. J., & Beck, S. (2011). Epigenome-wide association studies for common human diseases. *Nature Reviews Genetics*, *12*, 529–541.

Raney, B. J., Cline, M. S., Rosenbloom, K. R., Dreszer, T. R., Learned, K., Barber, G. P., et al. (2010). ENCODE whole-genome data in the UCSC genome browser (2011 update). *Nucleic Acids Research*, *39*, D871–D875.

Raychaudhuri, S., Plenge, R. M., Rossin, E. J., Ng, A. C., Purcell, S. M., Sklar, P., et al. (2009). Identifying relationships among genomic disease regions: Predicting genes at pathogenic SNP associations and rare deletions. *PLoS Genetics*, *5*, e1000534.

Richards, A. L., Jones, L., Moskvina, V., Kirov, G., Gejman, P. V., Levinson, D. F., et al. (2011). Schizophrenia susceptibility alleles are enriched for alleles that affect gene expression in adult human brain. *Molecular Psychiatry*, *17*, 193–201.

Robinson, J. T., Thorvaldsdóttir, H., Winckler, W., Guttman, M., Lander, E. S., Getz, G., et al. (2011). Integrative genomics viewer. *Nature Biotechnology*, *29*, 24–26.

Roeder, K., Devlin, B., & Wasserman, L. (2007). Improving power in genome-wide association studies: Weights tip the scale. *Genetic Epidemiology*, *31*, 741–747.

Rosenbloom, K. R., Dreszer, T. R., Long, J. C., Malladi, V. S., Sloan, C. A., Raney, B. J., et al. (2011). ENCODE whole-genome data in the UCSC Genome Browser: Update 2012. *Nucleic Acids Research*, *40*, D912–D917.

Saccone, S. F., Bierut, L. J., Chesler, E. J., Kalivas, P. W., Lerman, C., Saccone, N. L., et al. (2009). Supplementing high-density SNP microarrays for additional coverage of disease-related genes: Addiction as a paradigm. *PLoS One*, *4*, e5225.

Saccone, S. F., Bolze, R., Thomas, P., Quan, J., Mehta, G., Deelman, E., et al. (2010). SPOT: A web-based tool for using biological databases to prioritize SNPs after a genome-wide association study. *Nucleic Acids Research*, *38 Suppl*, W201–W209.

Saccone, N. L., Culverhouse, R. C., Schwantes-An, T. H., Cannon, D. S., Chen, X., Cichon, S., et al. (2010). Multiple independent loci at chromosome 15q25.1 affect smoking quantity: A meta-analysis and comparison with lung cancer and COPD. *PLoS Genetics*, *6*, e1001053.

Saccone, S. F., Hinrichs, A. L., Saccone, N. L., Chase, G. A., Konvicka, K., Madden, P. A., et al. (2007). Cholinergic nicotinic receptor genes implicated in a nicotine dependence association study targeting 348 candidate genes with 3713 SNPs. *Human Molecular Genetics*, *16*, 36–49.

Saccone, S. F., Quan, J., & Jones, J. P. (2012). BioQ: Tracing experimental origins in public genomic databases using a novel data provenance model. *Bioinformatics*, *28*, 1189–1191.

Saccone, S. F., Quan, J., Mehta, G., Bolze, R., Thomas, P., Deelman, E., et al. (2011). New tools and methods for direct programmatic access to the dbSNP relational database. *Nucleic Acids Research*, *39*, D901–D907.

Saccone, N. L., Saccone, S. F., Goate, A. M., Grucza, R. A., Hinrichs, A. L., Rice, J. P., et al. (2008). In search of causal variants: Refining disease association signals using cross-population contrasts. *BMC Genetics*, *9*, 58.

Saccone, S. F., Saccone, N. L., Swan, G. E., Madden, P. A., Goate, A. M., Rice, J. P., et al. (2008). Systematic biological prioritization after a genome-wide association study: An application to nicotine dependence. *Bioinformatics*, *24*, 1805–1811.

Saccone, N. L., Wang, J. C., Breslau, N., Johnson, E. O., Hatsukami, D., Saccone, S. F., et al. (2009). The CHRNA5-CHRNA3-CHRNB4 nicotinic receptor subunit gene cluster affects risk for nicotine dependence in African-Americans and in European-Americans. *Cancer Research*, *69*, 6848–6856.

Samuel Reich, E. (2011). Cancer trial errors revealed. *Nature*, *469*, 139–140.

Sayers, E. W., Barrett, T., Benson, D. A., Bolton, E., Bryant, S. H., Canese, K., et al. (2011). Database resources of the National Center for Biotechnology Information. *Nucleic Acids Research*, *40*, D13–D25.

Schaefer, C., Meier, A., Rost, B., & Bromberg, Y. (2012). SNPdbe: Constructing an nsSNP functional impacts database. *Bioinformatics*, *28*, 601–602.

Sherry, S. T., Ward, M. H., Kholodov, M., Baker, J., Phan, L., Smigielski, E. M., et al. (2001). dbSNP: The NCBI database of genetic variation. *Nucleic Acids Research*, *29*, 308–311.

Sherva, R., Wilhelmsen, K., Pomerleau, C. S., Chasse, S. A., Rice, J. P., Snedecor, S. M., et al. (2008). Association of a single nucleotide polymorphism in neuronal acetylcholine receptor subunit alpha 5 (CHRNA5) with smoking status and with 'pleasurable buzz' during early experimentation with smoking. *Addiction*, *103*, 1544–1552.

Smith, E. N., Koller, D. L., Panganiban, C., Szelinger, S., Zhang, P., Badner, J. A., et al. (2011). Genome-wide association of bipolar disorder suggests an enrichment of replicable associations in regions near genes. *PLoS Genetics*, *7*, e1002134.

Stein, L. (2001). Genome annotation: From sequence to biology. *Nature Reviews Genetics*, *2*, 493–503.

Stein, L. D., Mungall, C., Shu, S., Caudy, M., Mangone, M., Day, A., et al. (2002). The generic genome browser: A building block for a model organism system database. *Genome Research, 12*, 1599–1610.

Stevens, V. L., Bierut, L. J., Talbot, J. T., Wang, J. C., Sun, J., Hinrichs, A. L., et al. (2008). Nicotinic receptor gene variants influence susceptibility to heavy smoking. *Cancer Epidemiology, Biomarkers & Prevention, 17*, 3517–3525.

Stormo, G. D. (2011). An introduction to recognizing functional domains. *Current Protocols in Bioinformatics*, Chapter 2, Unit 2.1.

The Gene Ontology Consortium, (2011). The Gene Ontology: Enhancements for 2011. *Nucleic Acids Research, 40*, D559–D564.

The Wellcome Trust Case Control Consortium, (2007). Genome-wide association study of 14,000 cases of seven common diseases and 3,000 shared controls. *Nature, 447*, 661–678.

Thorgeirsson, T. E., Geller, F., Sulem, P., Rafnar, T., Wiste, A., Magnusson, K. P., et al. (2008). A variant associated with nicotine dependence, lung cancer and peripheral arterial disease. *Nature, 452*, 638–642.

Thorgeirsson, T. E., Gudbjartsson, D. F., Surakka, I., Vink, J. M., Amin, N., Geller, F., et al. (2010). Sequence variants at CHRNB3-CHRNA6 and CYP2A6 affect smoking behavior. *Nature Genetics, 42*, 448–453.

Voineagu, I., Wang, X., Johnston, P., Lowe, J. K., Tian, Y., Horvath, S., et al. (2011). Transcriptomic analysis of autistic brain reveals convergent molecular pathology. *Nature, 474*, 380–384.

Wang, J. C., Cruchaga, C., Saccone, N. L., Bertelsen, S., Liu, P., Budde, J. P., et al. (2009). Risk for nicotine dependence and lung cancer is conferred by mRNA expression levels and amino acid change in CHRNA5. *Human Molecular Genetics, 18*, 3125–3135.

Wang, K., Li, M., & Hakonarson, H. (2010). Analysing biological pathways in genome-wide association studies. *Nature Reviews Genetics, 11*, 843–854.

Wang, K., Zhang, H., Ma, D., Bucan, M., Glessner, J. T., Abrahams, B. S., et al. (2009). Common genetic variants on 5p14.1 associate with autism spectrum disorders. *Nature, 459*, 528–533.

Ward, L. D., & Kellis, M. (2012). HaploReg: A resource for exploring chromatin states, conservation, and regulatory motif alterations within sets of genetically linked variants. *Nucleic Acids Research, 40*, D930–D934.

Wegiel, J., Kuchna, I., Nowicki, K., Imaki, H., Marchi, E., Ma, S. Y., et al. (2010). The neuropathology of autism: Defects of neurogenesis and neuronal migration, and dysplastic changes. *Acta Neuropathologica, 119*, 755–770.

Weiss, R. B., Baker, T. B., Cannon, D. S., von Niederhausern, A., Dunn, D. M., Matsunami, N., et al. (2008). A candidate gene approach identifies the CHRNA5-A3-B4 region as a risk factor for age-dependent nicotine addiction. *PLoS Genetics, 4*, e1000125.

Westesson, O., Skinner, M., & Holmes, I. (2012). Visualizing next-generation sequencing data with JBrowse. *Briefings in Bioinformatics*, (in press).

Wingender, E. (2008). The TRANSFAC project as an example of framework technology that supports the analysis of genomic regulation. *Briefings in Bioinformatics, 9*, 326–332.

Wu, C. C., Huang, H. C., Juan, H. F., & Chen, S. T. (2004). GeneNetwork: An interactive tool for reconstruction of genetic networks using microarray data. *Bioinformatics, 20*, 3691–3693.

Yandell, M., Huff, C. D., Hu, H., Singleton, M., Moore, B., Xing, J., et al. (2011). A probabilistic disease-gene finder for personal genomes. *Genome Research, 21*, 1529–1542.

Yuan, H. Y., Chiou, J. J., Tseng, W. H., Liu, C. H., Liu, C. K., Lin, Y. J., et al. (2006). FASTSNP: An always up-to-date and extendable service for SNP function analysis and prioritization. *Nucleic Acids Research, 34*, W635–W641.

Zhang, J., Feuk, L., Duggan, G. E., Khaja, R., & Scherer, S. W. (2006). Development of bioinformatics resources for display and analysis of copy number and other structural variants in the human genome. *Cytogenetic and Genome Research, 115*, 205–214.
Zhao, J., Miles, A., Klyne, G., & Shotton, D. (2009). Linked data and provenance in biological data webs. *Briefings in Bioinformatics, 10*, 139–152.
Zhou, X., Maricque, B., Xie, M., Li, D., Sundaram, V., Martin, E. A., et al. (2011). The Human Epigenome Browser at Washington University. *Nature Methods, 8*, 989–990.

SUBJECT INDEX

Note: Page numbers followed by "*f*" indicate figures, and "*t*" indicate tables.

O

P

R

S

T

Z

CONTENTS OF RECENT VOLUMES

Volume 39

Volume 40

Volume 41

Volume 42

Volume 43

Volume 44

Volume 45

Volume 46

Volume 47

Volume 48

Volume 49

Volume 50

Volume 51

Volume 52

Volume 53

Volume 54

Volume 55

Volume 56

Volume 57

Volume 58

Volume 59

Volume 60

Volume 61

Volume 62

Volume 63

Volume 64

Volume 65

Volume 66

Volume 67

Volume 68

Volume 69

Volume 70

Volume 71

Volume 72

Volume 73

Volume 74

Volume 75

Volume 76

Volume 77

Volume 78

Volume 79

Volume 80

Volume 81

Volume 82

Volume 83

Volume 84

Volume 85

Volume 86

Volume 87

Volume 88

Volume 89

Volume 90

Volume 91

Volume 92

Volume 93

Volume 94

Volume 95

Volume 96

Volume 97

Volume 98

Volume 99

Volume 100

Volume 101

Volume 102

The Function and Mechanisms of Nurr1 Action in Midbrain Dopaminergic Neurons, from Development and Maintenance to Survival
Yu Luo

Monoclonal Antibodies as Novel Neurotherapeutic Agents in CNS Injury and Repair
Aruna Sharma and Hari Shanker Sharma

The Blood–Brain Barrier in Alzheimer's Disease: Novel Therapeutic Targets and Nanodrug delivery
Hari Shanker Sharma, Rudy J. Castellani, Mark A. Smith and Aruna Sharma

Neurovascular Aspects of Amyotrophic Lateral Sclerosis
Maria Carolina O. Rodrigues, Diana G. Hernandez-Ontiveros, Michael K. Louis, Alison E. Willing, Cesario V. Borlongan, Paul R. Sanberg, Júlio C. Voltarelli and Svitlana Garbuzova-Davis

Quercetin in Hypoxia-Induced Oxidative Stress: Novel Target for Neuroprotection
Anand Kumar Pandey, Ranjana Patnaik, Dafin F. Muresanu, Aruna Sharma and Hari Shanker Sharma

Environmental Conditions Modulate Neurotoxic Effects of Psychomotor Stimulant Drugs of Abuse
Eugene A. Kiyatkin and Hari Shanker Sharma

Central Nervous Tissue Damage after Hypoxia and Reperfusion in Conjunction with Cardiac Arrest and Cardiopulmonary Resuscitation: Mechanisms of Action and Possibilities for Mitigation
Lars Wiklund, Cecile Martijn, Adriana Miclescu, Egidijus Semenas, Sten Rubertsson and Hari Shanker Sharma

Interactions Between Opioids and Anabolic Androgenic Steroids: Implications for the Development of Addictive Behavior
Fred Nyberg and Mathias Hallberg

Neurotrophic Factors and Neurodegenerative Diseases: A Delivery Issue
Barbara Ruozi, Daniela Belletti, Lucia Bondioli, Alessandro De Vita, Flavio Forni, Maria Angela Vandelli and Giovanni Tosi

Neuroprotective Effects of Cerebrolysin, a Combination of Different Active Fragments of Neurotrophic Factors and Peptides on the Whole Body Hyperthermia-Induced Neurotoxicity: Modulatory Roles of Co-morbidity Factors and Nanoparticle Intoxication
Hari Shanker Sharma, Aruna Sharma, Herbert Mössler and Dafin Fior Muresanu

Alzheimer's Disease and Amyloid: Culprit or Coincidence?
Stephen D. Skaper

Vascular Endothelial Growth Factor and Other Angioglioneurins: Key Molecules in Brain Development and Restoration
José Vicente Lafuente, Naiara Ortuzar, Harkaitz Bengoetxea, Susana Bulnes and Enrike G. Argandoña

INDEX